SpringerBriefs in Climate Studies

SpringerBriefs in Climate Studies present concise summaries of cutting-edge research and practical applications. The series focuses on interdisciplinary aspects of Climate Science, including regional climate, climate monitoring and modeling, palaeoclimatology, as well as vulnerability, mitigation and adaptation to climate change. Featuring compact volumes of 50 to 125 pages (approx. 20,000–70,000 words), the series covers a range of content from professional to academic such as: a timely reports of state-of-the art analytical techniques, literature reviews, in-depth case studies, bridges between new research results, snapshots of hot and/or emerging topics.

Author Benefits: SpringerBriefs in Climate Studies allow authors to present their ideas and readers to absorb them with minimal time investment. Books in this series will be published as part of Springer's eBook collection, with millions of users worldwide. In addition, Briefs will be available for individual print and electronic purchase. SpringerBriefs books are characterized by fast, global electronic dissemination and standard publishing contracts. Books in the program will benefit from easy-to-use manuscript preparation and formatting guidelines, and expedited production schedules. Both solicited and unsolicited manuscripts are considered for publication in this series. Projects will be submitted to editorial review by editorial advisory boards and/or publishing editors. For a proposal document, please contact the Publisher.

Nisar Kannangara · Kalaiarasi Kandhan Sagunthala

Inequalities and Interventions

Autonomous Adaptation to Climate Change

Nisar Kannangara
Indian Institute for Human Settlements
(IIHS)
Bengaluru, Karnataka, India

Kalaiarasi Kandhan Sagunthala
Independent Researcher
Chennai, Tamil Nadu, India

ISSN 2213-784X ISSN 2213-7858 (electronic)
SpringerBriefs in Climate Studies
ISBN 978-981-96-5411-6 ISBN 978-981-96-5412-3 (eBook)
https://doi.org/10.1007/978-981-96-5412-3

© National Institute of Advanced Studies 2025

This work is subject to copyright. All rights are solely and exclusively licensed by the Publisher, whether the whole or part of the material is concerned, specifically the rights of translation, reprinting, reuse of illustrations, recitation, broadcasting, reproduction on microfilms or in any other physical way, and transmission or information storage and retrieval, electronic adaptation, computer software, or by similar or dissimilar methodology now known or hereafter developed.
The use of general descriptive names, registered names, trademarks, service marks, etc. in this publication does not imply, even in the absence of a specific statement, that such names are exempt from the relevant protective laws and regulations and therefore free for general use.
The publisher, the authors and the editors are safe to assume that the advice and information in this book are believed to be true and accurate at the date of publication. Neither the publisher nor the authors or the editors give a warranty, expressed or implied, with respect to the material contained herein or for any errors or omissions that may have been made. The publisher remains neutral with regard to jurisdictional claims in published maps and institutional affiliations.

This Springer imprint is published by the registered company Springer Nature Singapore Pte Ltd.
The registered company address is: 152 Beach Road, #21-01/04 Gateway East, Singapore 189721, Singapore

If disposing of this product, please recycle the paper.

Preface

This book tells ethnographic stories of spontaneous and intuitive responses and strategies of people and communities to navigate the impact of climate change. The ethnographies presented in this book are diverse since they are about people and communities inhabiting diverse terrains—coast, plain, and hills. The ethnographic insights from the field have wide significance in the production of academic knowledge and policy interventions on climate change.

The intuitive and spontaneous responses to coping with the shocks and tapping the opportunities induced by climate change termed as autonomous adaptation, is the central theme the book deals with. Autonomous adaptation wasn't a very popular theme in the climate change discourses. However, its significance has gradually been recognized in recent times. The Intergovernmental Panel for Climate Change (IPCC) recognized autonomous adaptation for the first time in assessment reports in 2007, and it gradually enhanced academic attention to autonomous adaptation to climate change in the last decades.

The realization among the scientists studying the changing climate, of the inevitability of prioritizing the adaptation strategies to compete with the climate crisis provides significance to the study of autonomous adaptation. As an adaptation that takes place without any planned interventions from the state or non-state actors, it has the quality to aid as well as disable the effects of state interventions. It is assumed that incorporating the autonomous adaptation strategies into the planned adaptation process could make a positive impact in eliminating the chances of both the planned adaptation and autonomous adaptation turning into maladaptation. Indeed, maladaptation is a stress multiplier that may lead climate change victims to an extremely vulnerable position.

There are also growing critiques on the dominance of the Western scientific epistemologies on climate change. These critics inevitably propose to look for alternative epistemologies—specifically the socio-economic-ecological-political nexus reflections from the grassroots, rather than relying only on the meteorological data and making new predictions and assessments on the future. This too has resulted in new field enquiries on the autonomous adaptation from different parts of the world. The existing studies on autonomous adaptation strategies—particularly explored how individuals and communities intuitively adapted to the livelihood crisis induced by the climate crisis viz livelihood diversification, changes in cropping patterns, migration, etc.

However, the study on autonomous adaptation seems an emerging field and its scope and significance in climate change studies are yet to be explored. Against this backdrop, the study presented in this book has investigated how autonomous adaptation interacts with the existing process of economic and social actions in a geographically diverse rural settings in Indian context. The geographical settings include the fisherfolk settlement on the Malabar coast, traditionally paddy-cultivating agrarian settlement on the plains, and a rural settlement that emerged depending on plantation crops on the hills. All three settings are chosen from Kozhikode in Kerala, identified as one of the districts that is most exposed to multiple climatic events in India. The investigation focused on the social and economic inequalities and observed how inequalities—specifically the interpersonal, and intra-personal inequalities intervene in the process of autonomous adaptation and vice versa. A detailed investigation of the complex and dynamic outcome of inequality and autonomous adaptation includes changes in labor relations, land relations, and relationships between communities, religious groups, and political ideologies.

The structure of the book is as follows; the Chap. 1 of this book "The Making of Autonomous Adaptation", provides conceptual frameworks on autonomous adaptation, its meanings, and relevance in climate change adaptation studies. Chapter 2 "Exposure to Climate Change", presents the district-wise mapping of India's exposure to climate change. This chapter draws heavily from the findings and analysis presented our article "Mapping India's exposure to climate change: a district level study", that we authored along with Kritika Singh and published in Current Science. This chapter provides the rationale for choosing Kozhikode as an ethnographic site to investigate autonomous adaptation. Chapter 3 "The Terrain and the Social Structure" provides a geographical and social background of the three rural settings, Melthura a fisherfolk settlement on the Malabar coast of Arabian Sea, Thamarakulam, paddy growing rural settlement in the plain, and Anappara a settlement developed on the plantation crops on the foothills of Western Ghats in Kozhikode district in Kerala. This chapter helps the readers to understand how the differences in terrain shape economic and social differences among the people in the same district, and the differential historical trajectory of the social and economic transformation in these settlements.

Chapter 4 "Sea Change in Melthura", delineates the ethnographic account of the way climate change has impacted the life of fisherfolk in the village, and various ways in which individual and different communities including the fishing and non-fishing communities responded, the act of inequality, and the inevitable outcome of the process. The specific process uncovered in the chapter includes semi-proletarianization, the movement towards a bigger harbor, and bigger boats, and the communal polarisation between the Hindu and Muslim fisherfolk. Chapter 5 "The Shifting Rains of Thamarakulam", explains how the rainfall variation and its subsequent impacts are making it difficult for the paddy cultivators in the village and their intuitive and spontaneous responses to the crisis. This chapter argued that one of

Preface

the outcomes of the interaction between the inequalities and autonomous adaptation to climate change led to the return of the tenancy in the paddy field which essentially provided benefits to the lower caste agricultural laborers in Thamarakulam. Chapter 6 "Sliding Plantations in Anappara" discusses the climate crisis in the plantation sector in the Anappara village and tells the story of the responses of planters and their laborers. The changes in relation and the process of reverse migration are the two important themes discussed in the chapter. Chapter 7 "External Interventions in Autonomous Adaptation", is the conclusion chapter of the book. In this chapter, we tried to discuss the issue that requires interventions, already existing local responses along with the possible future interventions and the authorities of implementation of the localised and context-specific interventions.

This book is an outcome of a multidisciplinary research project conducted at the Inequality and Human Development Programme of the National Institute of Advanced Studies, Bengaluru. The project began in February 2021 and completed in March 2024. We are grateful to Dr. Shailesh Nayak, Director of NIAS, and Prof. Narendar Pani, head of the Inequality and Human Development Programme and the Principal Investigator of the project, without their guidance and support this book would not be possible. We are also indebted to Kritika Singh for her inputs particularly shaping the second chapter and the faculty members and research staff at the Inequality and Human Development Programme, Dr. Anant Kamath, Dr. Chetan Choitani, Dr. Debosree Banerjee, Dr. Kshipra Jain, Dr. Mahima Upadhyay, Paul Thomas, Ishita Patil, and Ajith Kumar Babu, their comments and suggestion are immensely helped in the making of the book. We are also thankful to Srinivas Aithal, Head of Administration, NIAS, and Prashantha SC, Librarian at NIAS, for their kind support. Moreover, we also received immense support from the politicians, bureaucrats, community leaders, and people including fisherfolk, cultivators, laborers, and planters in Kozhikode, we are grateful to all of them. We are also grateful to Swati Meherishi and Bhagavati Murugasan of Springer for helping us to publish this work on time. The list is big, thanks to everyone who spent their time and support for this study.

We believe that this book has significance in the study of inequality and autonomous adaptation to climate change. Primarily it employed multidisciplinary research on autonomous adaptation. The first author of this book has trained in anthropology and ethnography, and the second author has trained in geography and climate science, besides the project also employed concepts and methods from economics in the theorization process. It is an inaugural book on the inequality and autonomous adaptation to climate change in Indian contexts which covers diverse geographical terrain and different social and economic settings. This would be beneficial for the students and researchers looking at climate change adaptation from diverse disciplinary perspectives including economics, anthropology, sociology, politics, and human geography. It will be insightful to the policymakers, activists, and journalists who wish to engage with climate crisis and adaptation strategies.

Bengaluru, India Nisar Kannangara
Chennai, India Kalaiarasi Kandhan Sagunthala

Contents

1	**The Making of Autonomous Adaptation**		1
	1.1 Looking to Adaptation		3
	1.2 Spontaneous and Autonomous Adaptation		7
2	**Exposure to Climate Change**		13
	2.1 Exposure		13
	2.2 Climate Variability and Climate Change		14
	2.3 Observed Impacts of Climate Change in India		16
		2.3.1 Temperature	16
		2.3.2 Heatwaves	17
		2.3.3 Rainfall	18
		2.3.4 Extreme Rainfall Events	19
		2.3.5 Floods	19
		2.3.6 Drought	20
		2.3.7 Tropical Cyclones	21
	2.4 Mapping Exposure		21
	2.5 Why Kozhikode?		25
3	**The Terrain and the Social Structure**		29
	3.1 Melthura: A Fishing Village in Kozhikode		30
		3.1.1 The Social Structure of Melthura	31
		3.1.2 Ways of Fishing in Melthura	32
	3.2 Thamarakulam—The Paddy Cultivating Midlands		34
		3.2.1 Social Structure of Thamarakulam	35
		3.2.2 The Jenmi System	35
		3.2.3 Land Reforms	35
		3.2.4 Green Revolution	36
	3.3 Anappara—The Hillslope in Kozhikode		37
		3.3.1 The Social Structure of Anappara	38
		3.3.2 The Labour Movement and the Arrival of Small Planters	39
	3.4 Transitions in Social Structure		39

4	**Sea Change in Melthura**		41
	4.1	Decline of Sardine, Mackerel, and the Loss of Fishing Beach	43
		4.1.1 Necessity of Increasing the Scale	44
	4.2	Seasonal In-Migration	46
	4.3	Proletarianization	46
	4.4	Consolidation of Identity Capital	47
	4.5	Transformation of the Social Structure	48
5	**The Shifting Rains of Thamarakulam**		51
	5.1	Changing Rice Fields	52
	5.2	The Floods of Artificial Recharge	54
	5.3	Responses	54
	5.4	The New Social Relations of Paddy Production	55
	5.5	Beyond the Rice Fields	57
6	**Sliding Plantations in Anappara**		63
	6.1	The Impact of Climate Change on Plantation Crops	64
		6.1.1 Jose—The Long-Standing Planter of Anappara	64
	6.2	The Plights of Seasonal In-Migrant Beekeepers	65
	6.3	Extreme Events in Anappara	66
	6.4	Economic Fragility of Anappara	67
	6.5	Changes in the Labour Relation	67
	6.6	Reversal of Migration	68
	6.7	Urbanization and Opportunities	69
7	**External Interventions in Autonomous Adaptation**		71
	7.1	Melthura—The Abandoned Fishing Beach	72
	7.2	Thamarakulam—The Flooded Rice Bowl	73
	7.3	Anappara—The Sliding Plantations	75
	7.4	Conclusion	76
References			77

List of Figures

Fig. 1.1	Triad of autonomous adaptation	10
Fig. 2.1	Districts exposed to long-term variability in temperature and rainfall	23
Fig. 2.2	Districts exposed to associated climatic events	25
Fig. 2.3	Districts most exposed to multiple climatic events	26
Fig. 5.1	Area, yield, and production of rice in Kozhikode between 1998 and 2020	61
Fig. 5.2	Economic considerations of social groups in Thamarakulam	61

Chapter 1
The Making of Autonomous Adaptation

As the 62-year-old Rajan[1] rowed his fibre boat over the shallow waters in what had been, barely a decade ago, one of the largest expanses of paddy fields in north Kerala, he was, perhaps unknowingly, responding to the multiple dimensions of change the region had seen. Rajan knew the land that was now a shallow lake very well, though he had never had a right to it. He did not own the land, nor was he a formal tenant. The only right he and his family had was to the boundaries of the paddy field, or more specifically, to the *Kaitha*[2] plant that grew on the boundaries and stemmed the flow of water into the fields. He and his father collected the leaves of the plant, exercising a right he and his scheduled caste, the Pulayas, had to these leaves. His mother and many Pulaya women used their expertise to make mats out of the collected Kaitha leaves.

He remembered his days as a child growing up near these paddy fields, where his father had worked as agricultural labour for the owner of the land, Moosa Haji. The child, Rajan, helped his father work on the paddy fields which turned into vegetable gardens during the summer. In the monsoons, when water from the nearby river entered the paddy fields, Rajan helped his father and other relatives to fish in the flooded fields. Later, in 1970s, when they received homestead land from the government as a result of the enactment of Kerala Land-Reform Amendment Act of 1969, the family moved to Ayyankandi, an area away from the mainstream locations in the village. Rajan went to the government school in his village but did not study beyond 9th standard. He chose instead to work as an assistant to a mason. Rajan was quick to pick up construction skills and became a mason at a very young age. He used this advantage to become a contractor in the village. He recalls the time when he had 10 labourers working for him. He was able to manage two to three construction contracts at a time. He got building construction contracts not only in his village but also in other villages and towns in the district. And then COVID-19 struck. As

---

[1] The names of all persons and villages in this study have been changed for reasons of confidentiality. The names of districts have been retained in order to help geographically locate the study.

[2] *Pandanus odorifer.*

construction activities came to a stop, he had nothing to offer his workers, and his work as a contractor plummeted.

It was here that climate change took a hand. It brought to his village a mix of floods and drought. Rajan had grown up in an environment where they fished during the floods and cultivated paddy and vegetables at other times. But now the flood waters did not recede, primarily due to an official intervention. The climate change-the changes in the monsoon rainfall, increasing temperature, and increasing extraction of ground water led to a depletion of groundwater levels, affecting the availability of drinking water to a nearby settlement. The official response to the need to raise groundwater levels was to flood the fields Rajan knew so well with water from a nearby irrigation canal. These fields were now perpetually flooded.

With the pandemic having introduced a break in his activities as a construction contractor, Rajan was looking for things to do. He saw an opportunity to try his hand in fish culture. He saw opportunities in the now permanently flooded paddy fields. The landowners, who were no longer able to cultivate the land, were open to putting it to other uses. In Rajan's words, "the paddy field was flooded in this part of the village, and no one was cultivating anything there. I approached a landowner, and he allowed me to do fish culture there. Later, other landowners also offered their land without really expecting very much in return."

Rajan's political activism—being a branch committee member of the Communist Party of India–Marxist (CPI-M), the ruling party in the state of Kerala—contributed to his awareness of government schemes that promoted fish culture. CPI-M is a dominant political party in his village, therefore his political loyalty helped him to get a subsidy easily from the Fisheries Department and bank loan from the Co-Operative Bank in the village to initiate his fish culture. His experiment with fish culture did not succeed at first. The species he was cultivating did not survive the onslaught of the creatures the river brought in when it flooded the paddy fields. An ecosystem filled with local fish species, wetland indigenous and migratory birds and other mammals were already thrived in the flooded paddy field was a major challenge to the fish culture. At the time of the fieldwork in 2023, Rajan said, "Now, I am doing both my contract work and fish culture together. On the first attempt, I didn't get anything in return from the fish culture. This time, I invested more and put up a solid net to protect the fish."

Rajan's story has a critical role in climate change, but it cannot be understood if climate change is seen in isolation. His experience of climate change was intertwined with that of the biological processes, the processes of worldliness, and the processes of plurality. The actions of biological processes keep life going, the actions of worldliness range from the fabrication of products that are durable enough for exchange to those that affect the status of the individual, and the actions of plurality cover the processes of interaction between persons that do not directly involve materials or things. The biological processes made their presence most strongly felt during the pandemic, but as a member of a scheduled caste, he knew what it was to stay alive in difficult economic circumstances. He learned to cope with the processes of worldliness when he dropped out of school and learned enough about the construction industry to become a labour contractor. And his recognition that no man is an island led to his finding a place for himself in the processes of pluralism by becoming a political activist. When landowners saw the perpetual flooding that resulted from climate change as a threat that prompted them to seek non-land options for earning their livelihood, Rajan saw the flooded fields as an opportunity for fish culture. Rajan's experience is a reminder that while isolating the effects of climate change from other socio-economic and political processes may make an interesting academic exercise, the experience of climate change can only be adequately explored when seen in its larger social context.

The experience of the contractor in the construction industry turned entrepreneur in fish culture also has another lesson for those examining climate change: the role of spontaneous responses to the phenomenon. Even as the discourse around climate change delves deep into issues of mitigation and adaptation, with prescriptions for both objectives, there are a large number of those who are affected by climate change who remain outside the influence of the discourse. Rajan did not associate the flooding of the fields with climate change, seeing it just as an act of government officials who caused the perpetual flooding to provide drinking water to a settlement. His spontaneous response to a local consequence of climate change was accompanied by a complete lack of awareness of the global phenomenon.

Rajan's experience of climate change interacting with a variety of social, economic, and political factors is, when seen in detail, unique. The specific combination of his socio-economic background, his relationship to land, the willingness of landowners to lease out land to him, and a variety of other aspects are unlikely to be repeated in identical terms elsewhere. In a broader sense, though, there would be many others who are responding spontaneously to other circumstances that have been influenced by climate change. When we recognize Rajan's experience as just one example of a spontaneous response to climate change, it raises several larger questions. How do individuals and groups react spontaneously to the consequences of climate change interacting with their wider socio-economic and political contexts? What are the consequences of these reactions? And what is the effect of these responses on social transformation and on climate change itself?

1.1 Looking to Adaptation

The questions raised by Rajan's experience, and indeed the larger questions of responses to climate change that are not mediated by expertise, call for a quick recollection of the course of adaptation to climate change. Adaptation has been understood as the adjustments made by the countries and communities in the processes, practices, and structures in response to the effects of climate change.[3] These adjustments could either contain the damages or aid in harnessing opportunities concerning the changing climate. Climate action strategies tend to depict adaptation to climate change as a set of adjustments formulated when the internal processes are challenged by external

---

[3] United Nations Framework Convention on Climate Change.

stressors i.e. physical effects of climate change.[4] The major area of focus of the international climate change adaptation discourse was to create funds and disseminate financial and technological support for the adaptation plans of the least and developing countries. The global adaptation strategies were pinned on the idea of adaptation being a linear process and largely incorporating sector-based responses. Based on the local arrangements adaptation measures range from building flood defences, setting up early warning systems for cyclones, and switching to drought-resistant crops, etc.,

In 1994 when the United Nations Framework Convention on Climate Change (UNFCCC) came into force, nation-states and other related stakeholders were asked to identify the options for adaptation and mitigation. When the framework organized goals under the Kyoto Protocol to mitigate the emissions of greenhouse gases based on scientific consensus, there was no similar concrete step towards adaptation. Climate impact assessments based on global models paved the path to understanding the need for adaptation. Countries gradually began communicating the findings of vulnerability and adaptation assessments at the national level.[5] In 2001 when the third assessment report of the International Panel on Climate Change (IPCC) was published, the overall stress on the need for adaptation increased. The major focus of the adaptation involved finance, capacity building, and technology transfer. Based on this, a series of programmes and mechanisms were launched by the international conventions on climate change to focus on adaptation.

In 2001, a work program on adaptation implementation in the Least Developed Countries (LDCs) was initiated with the preparation of the National Adaptation Programmes of Action (NAPAs).[6] These Programmes were based on the national circumstances of the Least Developed Countries (LDCs), with the list of adaptation activities or projects identified based on vulnerability or climate risk assessments. Further, the NAPA reports enabled the development of project proposals which eventually paved the way for the implementation of adaptation actions. The sector-based approach was also observed in these programmes and the priority sectors included agriculture, water resources, coastal zones, and disaster management.[7] In 2005, the Nairobi work programme was launched to assist all parties, especially the developing countries in understanding climate vulnerability and adaptation, and to shape decisions regarding measures for adaptation.

By 2010, global adaptation efforts were seeking ways to integrate climate change adaptation into socio-economic and environmental policies. As a result, the convention adopted the Cancun Adaptation Framework. Within this framework, countries agreed to consider vulnerable groups, communities, and ecosystems in their measures to adapt. Further, to concretize the adaptation action, at 16th the Conference of Parties (COPs) under the UNFCCC in 2010 an adaptation committee dedicated to promoting enhanced adaptation actions around the world was established. This was

---

[4] Scoones et al. (2023).
[5] UNFCCC (2013).
[6] National Adaptation Programmes of Action|UNFCCC.
[7] UNFCCC (2013).

1.1 Looking to Adaptation

one of the key institutions established by the convention to facilitate increased awareness of adaptation and to furnish support and guidance to the parties undertaking adaptation actions. The committee was involved in capacity-building activities and conducted workshops on ways to assess the impacts of climate vulnerability and shared knowledge on existing practices of adaptation.

Along with engaging in building awareness and improving the capacities of countries in adaptation, the framework worked channelling financial aid for the implementation of the adaptation programs through the creation of a special climate change fund under UNFCCC and an adaptation fund under the Kyoto Protocol. The developed countries that pledged to offer financial and technical assistance to the developing and under-developed countries had called for mainstreaming adaptation funds into the development funds.[8] Many climate change adaptation and mitigation initiatives also aided sustainable development. The coherence between development and climate funding was considered essential. This notion was adopted within the Cancun framework through the Green Climate Fund. The Green Climate Fund (GCF), established in 2010, started disbursing funds equally for both mitigation and adaptation. The adaptation projects under the mechanism encompassed themes related to health, food and water security, infrastructure, and ecosystem services. In 2015, the 21st Conference of Parties directed the GCF to facilitate the formulation and implementation of adaptation programmes in developing nations. Later in 2021, COP 26 at Glasgow pointed out the need for equal financial aid for adaptation as was provided for mitigation. Hence the Glasgow climate pact insisted on doubling the financial aid to developing countries for adaptation to the impact of climate change and for building climate resilience. As of October 2023, GCF was committed to 243 projects around the world[9] allocating nearly equal amounts for both mitigation and adaptation. To enhance the technology transfer a technology mechanism—Technology Executive Committee (TEC)—and a Climate Technology Centre and Network (CTCN) were established in 2010. These institutions allowed technology transfer between governments, the private sector, non-profit organizations, and research communities. As technology transfer involves the movement of know-how, tacit knowledge, or physical technology,[10] the process led to an increase in knowledge-sharing across the world.

Apart from building the capacities of nations the UNFCCC also recognized the need to "strengthen knowledge, technologies, practices and efforts of local communities and indigenous peoples related to addressing and responding to climate change".[11] In 2015, the Local Communities and Indigenous Peoples Platform came up as a forum to engage in the exchange of experiences and practices related to climate change adaptation and mitigation. The international bodies on climate change have been devoted to the idea that the nations and communities that are impacted by climate

---

[8] Smith et al. (2011).

[9] Fund (2023).

[10] Biagini et al. (2014).

[11] A Timeline of the Local Communities and Indigenous Peoples Platform|Local Communities and Indigenous Peoples Platform.

change need an improved awareness to adapt and are required to be equipped with technical assistance.

As an active participant in climate change negotiations, India has kept in touch with the international framework for both mitigation and adaptation. It ratified the UNFCCC in 1993 and has been vocal about the need to work towards climate mitigation and adaptation. The Indian government has officially recognized the challenge imposed by climate change on its economic growth, leading to the coming of the National Action Plan on Climate Change in 2008. This plan sought to identify measures to meet development objectives while simultaneously seeking the benefits of addressing climate change. The action plan consisted of eight missions out of which three focussed on mitigation, while the five others laid significant stress on adaptation enhancing energy efficiency, and natural resource conservation. These five missions focussed on themes related to agriculture, water, health, coastal zones, disaster management, Himalayan ecosystems, forestry, Capacity building, and Knowledge management. In the following year, considering the differential vulnerabilities at the sub-national level all the states in India were advised to devise a State Action Plan on Climate Change (SAPCC). These state plans were expected to be in alignment with the eight missions under the NAPCC. Since a large part of its population has been dependent on agriculture, there was a focus on this sector. Numerous policy initiatives were launched, including the National Food Security Mission, the National Mission for Sustainable Agriculture, and the Mission for Integrated Development of Horticulture to promote sustainable farming practices. This was made consistent with the emphasis on water resources management and the improvement of irrigation facilities.

In 2010 and 2011 the states, with the technical assistance of the development agencies, prepared action plans, some of which were based on elaborate assessments while others were not. The implementation of these plans faced several institutional barriers, with the plans proposed even falling short of resources for implementation. In 2015, to regulate the financial support to the projects proposed both at the national and state levels, the National Adaptation Fund for Climate Change was established. This fund enabled the implementation of some of the community-based projects in different states in the country. The major themes of the project once again involved agriculture, water management, and food security.

In an interesting step forward, in its Intended Nationally Determined Contribution detailing intended mitigation and adaptation strategies—, widely known as the INDC—India recognized the inter-state, inter-regional, and inter-group differences in the vulnerabilities to climate change. It began to consider the diversities in topography, climatic conditions, ecosystems as well as the local socio-economic structures. Even so, no clear methodologies were developed to capture those diversities during the impact assessments and implementation. For Instance, when hard adaptation measures like building a sea wall are planned the economic costs are estimated but how far the social and environmental costs are considered remains questionable. Even in terms of soft adaptation the effectiveness of community-level capacity building has rarely been analysed.[12]

In staying in step with the larger international framework for climate change India's approach to climate change adaptation has also been focussed on project formulation and implementation. The reduction of the challenge of adaptation to one of specific projects is not entirely justified in terms of the larger global discourse on climate change. In theory, adaptation is understood in broad enough terms with the IPCC defining adaptation as "the process of adjustment to actual or expected climate and its effects, to moderate harm or exploit beneficial opportunities."[13] In practice, this process has been seen as one of taking global knowledge to local situations. Adaptation strategies are then worked out at the global, regional, or national levels and then taken down to local communities in the form of specific projects. In the process, tracking adaptation becomes largely a matter of project formulation, implementation, and possibly, evaluation. It is also easier to justify the choice of a particular project when it seeks to benefit those who are vulnerable, not only to climate change but also to socioeconomic pressures.

India's support for this top-down approach to adaptation to climate change brought with it the question of who would pay for the projects and their implementation. Apart from the direct costs of the adaptation projects, there was the issue of the same absolute bill for adaptation being a greater burden for a poorer country than it was for the rich. This is linked to the costs of adaptation to the opportunity cost of mitigation if it requires India to forego development. Even as it operated within the larger international framework on climate change, India has tried to champion its cause and that of the cause of the poorer countries in response to sharing the financial burden of climate change.

The movement from the global to the local that has marked the perception of climate change has not been without its costs. The targets and objectives set by the top project are often too rigid for local situations. These adaptation strategies are often decontextualized in nature overlooking the political-economic structures and social processes that shape climate vulnerabilities and end up being technical and managerial solutions.[14]

1.2 Spontaneous and Autonomous Adaptation

When seen in terms of helping local populations cope with climate change impacts, the process of adaptation has a prominent place for planning by governments with access to expertise. There is expertise involved in determining how rapidly the earth is warming, how to limit that warming, and how to adapt to that part of it that is unavoidable. It is this expertise that is sought to be transferred to the vulnerable

---

[12] Smith et al. (2011).

[13] IPCC VI, p. 2898.

[14] Scoones et al. (2023).

who have been impacted by climate change. Adaptation is seen as largely a planned exercise that brings the benefits of expertise to the vulnerable. At the same time, there are a large number of those who have been impacted by climate change who have had little exposure to that expertise. Some are impacted by floods and landslides and have no access to the expertise that sees these events as the result of climate change. If the everyday is treated as the experience of ordinary people without the mediation of expertise, this can be seen as the everyday effects of climate change.

Individuals and groups respond spontaneously to the challenges posed by the everyday experience of climate change, as humans have done through the millennia of their existence. As these persons respond spontaneously to the threats, and opportunities, presented by climate change they are adapting to climate change, whether or not they recognize it as such. This has contributed to the distinction that is made between planned adaptation and autonomous adaptation. The IPCC has defined autonomous adaptation as "Adaptation in response to experienced climate and its effects, without planning explicitly or consciously focussed on addressing climate change."[15] It sees spontaneous adaptation as just another term for autonomous adaptation.

Acts of spontaneous adaptation to climate change do not occur in isolation. As individuals respond to manifestations of climate change, their actions affect those of others. The spontaneous actions of individuals and groups can interact with each other as they respond within the context of larger social processes. This interaction, which would include both coordination and conflict, results in a social process that has been modified by the spontaneous actions of individuals and groups responding to climate change. As these social outcomes are also not planned, they have a degree of autonomy to them. For convenience, we can distinguish between spontaneous adaptation and autonomous adaptation.

Spontaneous adaptation refers to the responses of individuals and groups to the everyday effects of climate change. Or to put it in terms consistent with the IPCC's definition of adaptation, spontaneous adaptation would be the process of adjustment to the everyday effects of climate change by individuals and groups, to moderate harm or exploit beneficial opportunities. Autonomous adaptation would be the cumulative effects of multiple spontaneous adaptations on social processes. In terms consistent with the IPCC's definition of adaptation, autonomous adaptation would be the adaptation of a social process to the everyday experience of climate change. While spontaneous adaptation would be seen in the actions of individuals and groups, autonomous adaptation would be seen in the social processes that result from these spontaneous adaptations and other actions they interact with.

Autonomous adaptation would, in this formulation, be influenced by the interaction of spontaneous adaptation with existing social processes. To the extent that a large part of the world's population can be expected to be unaware of the expertise of climate change, there is likely to be widespread spontaneous adaptation and the corresponding autonomous adaptation that has flown below the radar or analyses of climate change. These spontaneous adaptations, leading up to autonomous adaptations, form a significant part of the story of climate change. It tells us how people

---

[15] IPCC VI, p. 2898.

1.2 Spontaneous and Autonomous Adaptation

respond to the differences introduced in their everyday lives by the effects of climate change. Analyses of climate change entirely in terms of what experts think can then miss out on several aspects of the social impact of the phenomenon. A more complete picture of the social impact of climate change would need to also focus on the process of spontaneous adaptation leading up to autonomous adaptation.

The precise course this process takes would vary from situation to situation. The social processes would vary substantially across terrain, as would the nature of the everyday effects of climate change. While it would be difficult to reduce this diversity of experiences with everyday climate change to a single pattern, it is possible to view these diverse processes through a single analytical lens. The tool that is used in this study is that of the triad of transformation developed in Pani (2024). In its general form, the triad focuses on the continuous process of transformations generating differences that prompt responses which in turn generate further transformations. This process is driven by actions, which need not always be based on rational considerations. The actions generated by an irrational belief would have as much impact on social processes as a perfectly rational action. The triad thus helps us track processes as they occur rather than follow rational expectations of how they should occur. It would allow us to track all spontaneous responses to climate change, whether or not they are rational. We could then rationally follow their interaction with each other and with existing social processes, leading to transformations that have been arrived at autonomously.

We can enter the process represented in Fig. 1.1 at the point of social transformation that has been generated by climate change. These changes can be perceived differently by those who are affected by them. Some may see opportunities in a flooded rice field while others may only see the destruction of a traditional economic activity. Yet others may be totally indifferent to the change. These perceptions would influence the spontaneous adaptation to these differences. The spontaneous reaction of those who perceive an opportunity would be to invest in that opportunity; the perpetually flooded rice field would become a potential site for fish culture. Those who see it as the destruction of a traditional economic activity would respond by moving out of agriculture and perhaps even migrating out of the village. There would be others who would not be affected by the change and may choose not to respond. It is also possible that a common cause could be found among those responding spontaneously to the change. Those who see the flooded fields as an opportunity for fish culture could get together and demand policies that would suit this activity, just as those abandoning agriculture could seek support for non-agricultural activities. The result of this interaction of spontaneous adaptation with existing social processes would transform social processes in a way that could be seen as autonomous adaptation.

Mapping the interaction between spontaneous and autonomous adaptation through the lens of the triad of transformation calls for multiple empirical snapshots. The first snapshot would give us a picture of the terrain and the socioeconomic structures that have historically been associated with it. It would trace the socioeconomic relations of fishing communities in a coastal terrain, of agrarian communities in the plains, and of local and migrant communities in the hills. These snapshots would

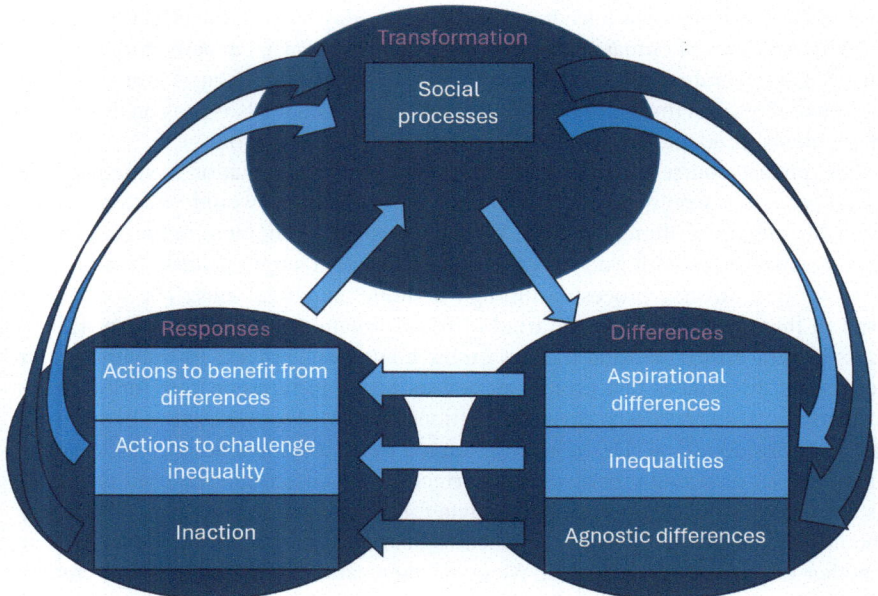

Fig. 1.1 Triad of autonomous adaptation

provide insights into the relationship between the individual, the groups, and the social processes. The detail of these snapshots would be enhanced by the insights of ethnography even as a primary survey would help distinguish between the exception and the rule.

The pictures of the human geography of specific terrains provide the backdrop against which we can picture the effects of climate change. When seen through the lens of the triad of transformation these effects do not occur in isolation but are intertwined with larger processes of social transformation. The effect of climate change perpetually flooded paddy fields amid a social transformation where farmers were moving out of agriculture, and into one where subsistence agriculture was the only option. In the former case, it would be no surprise if farmers could take the climate-related setback on their paddy fields in their stride, while in the latter case, it would make the farmers extremely vulnerable to climate change.

Since the focus of this study is on spontaneous and autonomous adaptation the snapshots would focus on the everyday experience of climate change that is the effects of climate change as they are felt by individuals and groups without the mediation of experts. The spontaneous responses to these everyday effects would constitute spontaneous adaptation. These spontaneous adaptations could vary across individuals, sometimes consistent with each other and sometimes contradictory. The consistency would not just result from similar responses. There could be cases where the spontaneous adaptation of one person is aided by a different spontaneous reaction of another person. The spontaneous adaptation of a farmer migrating away from a

1.2 Spontaneous and Autonomous Adaptation

farm that is now a perpetually flooded field would be consistent with the spontaneous adaptation of an individual seeking to carry out fish culture. Spontaneous adaptation of several individuals could also coincide. Perpetual flooding of paddy fields could lead several individuals to explore fish culture, thereby creating a group with common interests. There would then be the spontaneous response of a group to the everyday experience of climate change. Spontaneous responses could also be contradictory, as when one farmer would like to drain a flooded field while the other would like to create a permanent pond for fish culture. The cumulative effects of all these climate-related spontaneous actions on social processes would provide a picture of autonomous adaptation. The empirical basis for the picture of autonomous adaptation would be a mix of ethnography, large data on people and climate, and primary surveys.

When seen from this perspective autonomous adaptation is based on a climate-related, rather than socioeconomic, conception of the vulnerable. The multiple spontaneous acts of adaptation, influencing specific social processes that lead to autonomous adaptation, are not only those who are vulnerable in socioeconomic terms. The spontaneous actions can come from dominant groups in the village as well. Spontaneous adaptation is in response to the everyday effects of climate change, no matter who is affected. It would include the socially vulnerable as well as the socially dominant, those who would like to limit the harm caused by climate change as well as those who see it as an opportunity. The vulnerable would then be the community as a whole, depending on its exposure to climate change.

Chapter 2
Exposure to Climate Change

When the experience of climate change was regarded as a relational process, the dynamics in the variation gained prominence. The key determinants of climate risk were seen to involve exposure, vulnerability, and hazard. All three of these concepts looked at the adverse impacts of climate change on human systems as well as ecosystems. Where vulnerability looked into the susceptibility to be adversely affected by climate change, exposure denoted the spatial element, delving into the presence of people, livelihoods, and ecosystems in the settings which could be adversely affected and while the hazard deals with the possibilities in the occurrence of climate events or their long-term trends. Vulnerability assessment is one of the key tools used in planning climate change adaptation strategies all over the world. The Climate Change Vulnerability Assessments (CCVAs) focused on identifying the risks from the direct and indirect climate change impacts. These assessments also included the concepts of exposure, sensitivity, and adaptive capacity. But later the concept of exposure was delinked from vulnerability and became a pre-requisite to understanding vulnerability. Climate change adaptation planning was based on the internal processes affected by the external physical impacts, so the concept of exposure came forth as the means to study the physical impacts of climate change.

2.1 Exposure

Exposure deals with the climate stress levels of the unit of study. The utility of the concept of exposure varied over time within the discourses related to climate change mitigation and adaptation. When global climate change was fixated on mitigation, exposure was used more in terms of the health and economic consequences (Volatility in the fuel price) of climate change effects.[1] Later when the concept of adaptation

---

[1] Banuri et al. (2001).

gained importance the concept of exposure was part of the vulnerability component of the climate change debate. However, the exposure with the all-around experience of the physical impacts of climate change on a system. The third assessment report of the Intergovernmental Panel on Climate Change (IPCC),[2] defined exposure as 'the nature and degree to which a system is exposed to significant climatic variations.' But later in the Fifth assessment report[3] of the panel the definition was elaborated but stuck to just the adverse effects of climate change in the sense it defined exposure as the "Presence of people; livelihoods; species or ecosystems; environmental functions, services, and resources; infrastructure; or economic, social, or cultural assets in places and settings that could be adversely affected".

The effects of climate change did not always have adverse effects, some of them provided opportunities, For instance, the receding snow-line in the Nordic countries affected the security provided in terms of cross-country skiing yet the temperature rise increased the growing season of crops thus their production. So going by the initial definition where the exposure of humans and related systems at a particular place to variations in climate a more holistic understanding of the experience of climate change at the local level. The variations in the climate differ locally across varied geographies. For instance, floods in the hills might have completely varied impacts and responses. Vulnerability is the other major measure of climate risk. But as it just deals only with the internal property of a system, vulnerability might not be a sufficient measure as a system that is exposed may not be vulnerable, but to be vulnerable a system needs to be exposed to climate change. Exposure has been evaluated using the biophysical indicators or interest variables. The interest variable includes a group of variables denoted as the climate change determinant stressor.[4] On a similar note, IPCC came up with the Climate Impact Drivers (CID), which indicate the physical climatic conditions that affect humans or ecosystems. These drivers include the conditions of the physical climate system such as the means, events, and extremes.

2.2 Climate Variability and Climate Change

Human systems and ecosystems experience both variability in climate and climate change. Climate variability involves short-term fluctuations in the weather, while climate change is the change in the long-term averages over decades or longer. The climate varies over the seasons, the climate variability is noticeable with some summers colder than or warmer than others. When the climate fluctuates on a yearly basis or below the long-term average value it is known as climate variability. It also included events like El Niño and La Niña events. Climate variability is observed to

---

[2] Banuri et al. (2001).
[3] Field et al. (2014).
[4] Luers et al. (2003).

2.2 Climate Variability and Climate Change

be an effect of natural and also periodic changes in the circulation of the air and ocean, volcanic eruptions, and other factors.

The average weather of a region in terms of mean and variability of relevant quantities over a period is known as climate. Any systematic long-term change in the statistics of the climate variables over several decades or longer is characterized as climate change. The changing climate can be attributed to both natural and human-induced processes.[5] The range of the period varies between months and millions of years. However, the traditional average period considered by the World Meteorological Organization (WMO) is 30 years. The common surface variables that are considered in understanding the climate are temperature, precipitation, and wind.[6] Scientists use average weather conditions over 30-year time intervals to track the changing climate. These 30-year averages are called climatological normals. The climate normals are used reference to represent or determine the climate and its significant changes at a particular location as the 30 years of data is long enough to calculate an average that is not influenced by year-to-year climate variability.

The natural processes include changes in solar emission or changes in the earth's orbit, while the anthropogenic factors involve the increase in the emission of heat-trapping Greenhouse gases. This trapped heat in the atmosphere ends up warming the globe, the phenomenon popularly known as Global Warming.[7] The resultant effect of an increase in the global surface temperature was observed to pave the way for a series of compounding effects. The variability in the surface temperature characterises climate change. The warming climate increases evaporation which in turn results in changing the overall magnitude of the precipitation.[8] This variability in rainfall has its impacts on the hydrological processes thus determining the availability of the water and can exacerbate hydro-meteorological disasters.[9]

The state of the climate change is established through the set of indicators. The surface temperature acts as a key indicator of climate change as it increases quasi-linearly with cumulative greenhouse gas emissions.[10] According to the World Meteorological Organization report on the status of global climate in 2015, Temperature and precipitation were referred to as two of the key indicators among the other on greenhouse gases, sea-levels cryosphere.[11] In addition to the key indicators climate change is also associated with the type, frequency, duration, and intensity of events such as heat waves, cold waves, cyclones, floods, and droughts. The occurrence of Climate extremes corresponds to the fact that when the value of a climate variable breaches the thresholds either lower or upper ends. The climate events that are not extremes can still lead to extreme impacts by crossing a threshold in a particular social, ecological, or physical system or by occurring around the same time simultaneously with

---

[5] Buonocore (2024).
[6] IPCC, IPCC Glossary Search.
[7] Met Office.
[8] United States Environment Protection Agency (2022).
[9] Irwandi et al. (2023).
[10] Sanjay et al. (2020).
[11] Global Climate Observing System (GCOS) (2016).

another climate event. Like climate change phenomena climate extremes can be a result of both the natural and anthropogenic factors but the changing climate changes the frequency, intensity, spatial extent, duration, and timing of climate extremes.[12]

2.3 Observed Impacts of Climate Change in India

There are variations in how different regions around the world face the impacts of climate change. So understanding the regional differences in exposure to varied climatic events forms an essential step. Exposure to climate change involves both long-term and short-term events. According to the IPCC's sixth assessment report,[13] the Indian sub-continent is expected to experience a further rise in the sea level, erratic monsoon rainfall, severe heat waves, and tropical cyclones in the coming years. Similarly, the Ministry of Earth Sciences reported variations in the surface temperature, monsoon precipitation, droughts, floods, sea-level rise in the North Indian Ocean, Cyclonic storms, and Himalayan cryosphere as the key climatic indicators for the Indian sub-continent.[14]

2.3.1 Temperature

The increase in temperature all around the world was attributed to the emission of greenhouse gases and also the change in land use and land cover. Between 1901 and 2015, there has been a significant positive trend in the temperature in most of the country. The trend ranged between 0.5 and 1.5 °C in mean annual temperature with a 95% significance level.[15] The historical simulations of IITM-ESM between 1901 and 2018 also came up with a similar observation with an increase of 0.7 °C in the average temperature. North and the north-western part of the country experienced significant warming between 1951 and 2014.

Change temperature patterns in the country were observed through the patterns in both annual and seasonal variations in maximum, minimum, and mean temperature. The Seasonal divisions include winter (January–February), pre-monsoon (March–April–May), monsoon (June–July–August–September), and post-monsoon (October–November–December). The long-term trends in the surface temperature between 1980 and 2020, indicate a significant increase during Pre and Post monsoon seasons in the northwest, northeast, and north-central India. Per decade the temperature variations ranged between 0.1 and 0.3 °C in pre-monsoon and between 0.2 and

---

[12] Seneviratne et al. (2012).
[13] IPCC (2021).
[14] Krishnan et al. (2020).
[15] Indian Meteorological Department (IMD) (2016).

0.4 °C in these regions. The post-monsoon temperature had shown a positive trend in almost all over the country.[16]

The agricultural development of the country depends on the variations in the trends of temperature. Notable levels of variability in the maximum as well as the minimum temperature had been observed in different agro-climatic zones in India. Increase in the annual maximum temperature in the parts of Himachal Pradesh, erstwhile Jammu and Kashmir, and Uttarakhand, in all North-eastern states, Odisha, Jharkhand, and Chhattisgarh, in Maharashtra, and parts of Southern states of Karnataka, Telangana, Andhra Pradesh, Kerala, and Tamil Nadu. Significant reduction in maximum temperatures has taken place in a few agro-climatic zones of Rajasthan, Uttar Pradesh, and Tamil Nadu. A significant rise in annual minimum temperatures was seen in most parts of the Northern and Central states, in the Eastern States of Bihar, West Bengal, and Sikkim, and parts of the Southern states of Karnataka, Kerala, and Tamil Nadu.[17]

2.3.2 Heatwaves

A prolonged period of excessive heat with abnormally high temperatures is defined as a heatwave. Quantitatively, based on the temperature thresholds experienced over a particular region the heat wave event is declared. In general, an event is considered a heat wave when the maximum temperature at a station reaches at least 40 °C in plains and 30 °C in hills. However, the normal and actual temperatures of different locations vary over seasons. So, when there is a 4.5–6.4 °C departure from the normal temperature it is marked by the IMD as a heat wave when this departure from the normal temperature is >6.4 °C then it is considered a severe heat wave. In terms of actual temperature, when the actual temperature of a location touches 45 °C or more it is known as a heat wave and if it is more than or equal to 47 °C it is a severe heat wave. Heat waves are very commonly experienced during summer between April and June in India.[18]

The abnormally high temperatures lead to physiological stress in the human body sometimes even causing death. A study on the deaths due to heat waves in India between 1971 and 2010 revealed the country had witnessed relatively the highest number of deaths between 2001 and 2010. The projections on temperature variability also suggest that the average temperature in India will approximately rise by 4.4 °C at the end of the twenty-first century. The frequency of warm days and nights has been also projected to increase by 50 and 70% respectively. The mean duration of the heatwave was also projected to double. The rising surface temperature and humidity were observed to amplify the heat stress all over India especially over the Indo-Gangetic and Indus River basins. On average two heat wave days during the summer season were witnessed in most parts of the country between 1961 and

---

[16] Krishnan et al. (2020).
[17] Chattopadhyay et al. (2019).
[18] Indian Meteorological Department (IMD) (2020).

2010 except for north-east and peninsular India.[19] This zone was termed the Core HW zone which includes the states of Rajasthan, Uttar Pradesh, Gujarat, Madhya Pradesh, Chhattisgarh, Bihar, Jharkhand, West Bengal, Odisha, Punjab, Himachal Pradesh, Uttarakhand, Delhi, Haryana, and Telangana. The core zone also includes the meteorological subdivisions of Marathwada, Vidarbha, Madhya Maharashtra, and coastal Andhra Pradesh. The most frequent bouts of heat waves were witnessed in parts of India such as East Rajasthan, Eastern and Western Uttar Pradesh, Eastern and Western Madhya Pradesh, Vidarbha, Odisha, Jammu and Kashmir, Himachal Pradesh, Punjab, West Rajasthan and Coastal Andhra Pradesh between 2010 and 2016. But around the same period, there were less frequent or absence of heat waves in Rayalaseema and some parts of Karnataka and Tamil Nadu.[20]

2.3.3 Rainfall

Along with temperature, the variability in precipitation is the other notable consequence of climate change. The types of precipitation include drizzle, rain, sleet, snow, and hail. Rainfall is the most common form of precipitation witnessed widely across India. The Indian monsoons result from the interaction between the land, ocean, and atmosphere. Thus the variations in the patterns become an important indicator of climate change. The patterns of variations in India are studied both on an annual and seasonal basis. The seasons of rainfall are most widely classified as pre-monsoon (MAM), monsoon (JJAS), post-monsoon (OND), and winter (JF). The Indian summer monsoon also known as the South West Monsoon brings rainfall in most parts between June to September and contributes around 70% of the annual average rainfall. But few parts of India also depend heavily on the North-East Monsoon or the Post monsoon. The states in the southern peninsular region especially situated on the eastern Coromandel coastline receive a large proportion of their rainfall during this monsoon between October and December. Likewise, the western side of the Himalayan region receives its rainfall from the winter monsoon induced by the western disturbances.[21]

The southwest monsoon was observed to have declined by 6% between 1951 and 2015. A notable proportion of this decline was felt in the Indo-Gangetic plains and the Western Ghats region. The weakening of the southwest monsoon observed post-1950s has been connected to rapid warming of the equatorial Indian Ocean and the variations in the sea surface temperature along with the anthropogenic aerosol forcing and land use land cover change at the regional level. The models predicting climate change at the regional scale such as CMIP5 had projected an increase in variability of the mean annual and monsoon precipitation as well as extreme rainfall events by the end of the twenty-first century. An analysis based on high-resolution gridded satellite

---

[19] Krishnan et al. (2020).

[20] Krishnan et al.

[21] Krishnan et al.

2.3 Observed Impacts of Climate Change in India

data for the period of 1901–2015 suggests a significant decline in the annual and seasonal rainfall over parts of Uttar Pradesh, Madhya Pradesh, Chattisgarh, Kerala, and the Western Ghats region, and some parts of the northeastern states. On the other hand, rainfall over parts of the country such as Goa, Jammu Kashmir, Gujarat, and the Konkan coast as well as the east coast shows a significant increasing trend. There are several extremes associated with the variability in rainfall. The increasing trends could cause floods, and a significant decrease in rainfall could cause droughts.[22]

2.3.4 Extreme Rainfall Events

While the mean annual and summer monsoon rainfall witnessed a significant decrease between 1951 and 2015, locally the frequency of heavy rainfall events has increased in India. A rainfall event is qualified as a heavy rainfall event when the 24-h rainfall ranges between 64.5 and 115.5 mm and when it exceeds or is equal to 204.5 mm per day is denoted as an extreme rainfall event in India. These heavy rainfall events can be triggered by cloudbursts, thunderstorms, or even monsoon lows. Generally, the impacts of extreme rainfall events are not limited to the point station where it is recorded but may have wider impacts in the form of flood events, health, water contamination, and other societal problems.[23] Floods related to the extreme rainfall events alone led to economic losses of 3 billion dollars per year. In the period between 1901 and 2010, an increase in high-intensity rainfall has been observed over the Indian subcontinent. The other findings also indicated the spatial coverage of the occurrence of extreme rainfall events has increased between 1961 and 2015 compared to the 6 decades before that. A significant three-fold increase in extreme rainfall events has been observed over central India.[24]

2.3.5 Floods

Floods occur when a water body starts overflowing due to the accumulation of inflow and submerges the plains which are usually not under water. Floods can take place from a few hours to days. Floods may result from various factors such as heavy rainfall events, duration of a storm, snowmelt, nature of the drainage basin, etc. Based on the causes and the type of location and terrain the floods can be classified into riverine, flash, coastal, urban, and pluvial flooding. India is highly vulnerable to floods. More than 40 million hectares in the country are identified as flood-prone. Every year,

---

[22] Krishnan and Dhara (2020).
[23] Aune et al. (2021).
[24] Roxy et al. (2017).

on average 75 lakh hectares of land experience flood. The regional attributes of a location define the characteristics of a flood.[25]

However, it is not easy to distinguish between the climate and non-climate drivers of climate change. Different studies have suggested how the impact of human-induced climate change on the hydrological cycle would influence floods. Studies suggest that the trend in heavy rainfall leading to local flooding in some parts of the country would increase and also increased frequency of the heavy rainfall on daily timescales would enhance the flood risk in India. Since 2000 the major urban centres such as Mumbai, Bengaluru, Chennai, Kolkata, and Ahmedabad have witnessed increased frequency of urban floods.[26] On the other hand, the major river basins like Brahmaputra, the Indus Basin, Narmada-Tapi, Godavari, Ganga, Mahanadi, etc. had experienced an increase in flooding. The abrupt change in the frequency of floods in India has been observed since 2005. This was reflected in a study on flood events between 1970 and 2019, post-2005 the number of extreme flood events per year has increased along with the spatial extent. In the year 2005 itself, the country witnessed 18 extreme floods, and around 69 districts were declared affected. The year 2019 in comparison registered 16 extreme floods in total affecting around 151 districts.[27]

2.3.6 Drought

Drought is the climate extreme characterized by the lack of moisture due to a reduction in the amount of rainfall over an extended period. It is classified into four types viz., meteorological drought when there is a deficit of seasonal precipitation over a long period; hydrological drought, as an inadequate stream flows, groundwater level, or water storage; agricultural drought, as a deficit in soil moisture; and socio-economic drought when an abnormal change in the water supply and demand affects the socio-economic functioning of the society. It has been observed that 68% of the land area in the country is prone to droughts of various degrees.[28] Drought can be caused by various factors insufficient rainfall, high evapotranspiration, and over-exploitation of water resources but usually, rainfall is considered the most important factor in the indices measuring drought. The drought indices include Percentage of Normal Precipitation (PNP), Standard Precipitation Index (SPI), Standard Precipitation Evapotranspiration Index (SPEI), etc. Among them, the SPI and SPEI are the most widely used.[29]

The significant variability and weakening of the summer monsoon is observed in India. It has been observed that droughts in the country increased between 1951 and 2015. During this time areas affected by drought increased by 1.3% per decade. The

---

[25] Floods|NDMA, GoI.
[26] Singh et al. (2023).
[27] Mohanty (2020).
[28] Rawat et al. (2022).
[29] Mujumdar et al. (2020).

regions in central India, the southwest coast, the southern peninsula, and northeast India on average experienced more than two droughts per decade between 1951 and 2015. The projections from the IPCC also suggest that by the end of the twenty-first century, the frequency of agricultural and meteorological droughts is likely to witness an increase in the dry regions due to climate change. These projections noted that the increased variability in the Indian monsoon rainfall could impact the intensity and frequency of droughts in India.[30]

2.3.7 Tropical Cyclones

Tropical cyclones are considered a multi-hazard phenomenon as they include varied climate events like heavy rainfall, gale wind, and storm surges during the landfall. Tropical cyclones are classified by the Indian Meteorological Department from low-pressure (<31 kmph) to super-cyclonic storms (≥ 222 kmph) based on maximum sustained wind speed.[31] Both the Arabian Sea and the Bay of Bengal in the North Indian Ocean are exposed to tropical cyclones. The post-monsoon (October–December) season in India is usually the peak season for cyclones in India. Though a decline in the frequency of the annual cyclones in India since 1951. The frequency of severe cyclonic storms has been observed to have increased between 2000 and 2018.[32]

A projected rise in the intensity of tropical cyclones over the North Indian Ocean during the twenty-first century has been indicated in the climate simulation models. In terms of frequency of the tropical cyclones, significant decline in the number of cyclones per decade in the North Indian Ocean during 1891–2018 and 1951–2018. However different trends in the frequency have been observed in the west and east coasts of the country. On the eastern coast of the Bay of Bengal, though a rising trend has been marked especially between November and May, the frequency of the cyclonic and severe cyclonic storms has declined significantly between 1951 and 2018. While on the western coast (Arabian Sea) upward trend of cyclones of similar intensity during the post-monsoon season has been observed.[33]

2.4 Mapping Exposure

Based on the observed and projected trends we identified temperature, precipitation, floods, heatwaves, droughts extreme rainfall events, and cyclones as the vital climate Impact Drivers (CIDs) relevant to the Indian sub-continent to map the most exposed

---

[30] Prathipati et al. (2019).
[31] Indian Meteorological Department (IMD) (2021).
[32] Vellore et al. (2020).
[33] Vellore et al.

district in India. According to Ian Burton, there are three levels of climate indicators, The first type consists of single variables (Temperature, Precipitation, etc.) and associated extremes of those single variables (heat waves, cold waves, droughts, and floods); the Second type is of complex climate phenomena like cyclones, tornadoes, etc. and the third type is the most complex which involves the measurement of various factors such as air pollution.[34] Though the climate indicators are of varied nature but are interconnected in a way one can trigger the other. So to capture the compounding as well as cascading effects of climate change[35] for instance the continuous precipitation at a location could lead to floods, landslides, and various other impacts, In our study, we mapped the district-level exposure to multiple climatic events in three stages. Where when a district is marked to have been exposed to the most number of indicators in the first stage has been considered and carried over to analyse the exposure to climate extremes. Though this exercise would not directly represent the results of the compounding effects of climate change, it would be able to present the probability that a single location could be exposed the most to varied climate events.

In the first stage, we focussed on, single variables such as Temperature and Precipitation, the extreme variability in them could trigger other associated extremes. So in the following stage, we looked into the climate extreme, where we focussed on the trends in frequency and intensity of the extremes. The second stage involved climate extremes such as floods, heatwaves and droughts that are directly related to the extreme changes in the key climate indicators enquired in the first stage, the third stage involved complex events like cyclones and extreme rainfall events as their occurrence also involved components like oceanic circulation and local convection rates beyond the extreme variations in warming and precipitation.

As the variation of different climatic parameters is measured using different units we developed a matrix, when a district was found to have been exposed to the effects of a particular climatic parameter with a significant (increasing or decreasing) trend then those districts were assigned with the value 1, and the districts with no significant trend were marked with the value 0. This matrix was built based on the data from the secondary literature on district-level exposure to different climatic events. The matrix has three stages based on the characteristics of the climate indicator assessed. The three stages, relevant secondary data sources, and the results are as follows:

Stage 1: Key Indicators

This stage of the matrix considered the significant variability in the long-term trends of the two parameters as Temperature (Maximum and Minimum) and Rainfall. These two are known as the key indicators, the variability in them could legitimize the occurrence of associated climatic events. Moreover, fluctuations in the temperature and precipitation could alter and have a wider impact on the hydrological and environmental processes.

The criteria used while mapping the parameters are:

---

[34] Burton (1997).
[35] Cutter (2018).

2.4 Mapping Exposure

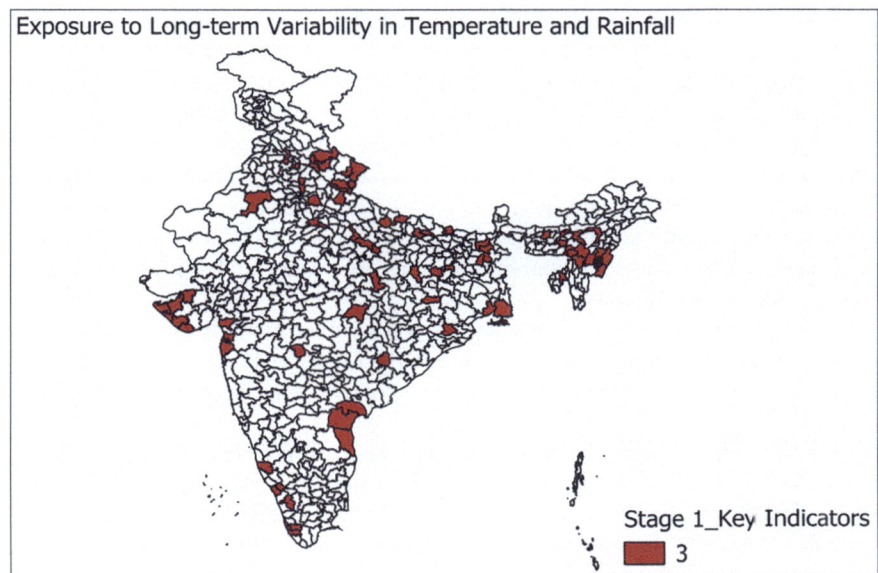

Fig. 2.1 Districts exposed to long-term variability in temperature and rainfall

Rainfall: District-wise rainfall variability analysis for various states in the country published by the Indian Meteorological Department (IMD) was extracted. These reports were noted to have used the daily rainfall data between 1989 and 2018. In the matrix, we noted the districts showing increasing or decreasing trends at a 95% significance level.[36]

Temperature: The climate research and services under IMD, Pune had mapped the long-term district-level trends in the annual average maximum and minimum temperature. These maps represented data between the periods 1901–2015.[37] The districts showing significant positive departure or negative departure (at a 95% significance level) from the average annual maximum/minimum temperature from the reference period 1981–2010 were marked in our matrix.

The results of the first stage revealed that 78 districts (Fig. 2.1) in the country have been exposed to the all three parameters i.e. Rainfall, Maximum temperature, and Minimum temperature.

Stage 2: Associated Climatic Events

As already noted we are looking to enquire about the compounding and cascading exposure to climate change, in stage two we focussed on the 78 districts identified in the previous stage. This stage considers Floods, Droughts, and Heat waves. The conditions used while considering these events are as follows:

---

[36] Indian Meteorological Department (2021).

[37] Indian Meteorological Department (2024).

Floods: The district-wise data on the magnitude of the incidence of floods and their intensity were obtained from the web portal developed by the Global Facility for Disaster Reduction and Recovery (GFDRR).[38] The portal represented the intensity, frequency, and susceptibility of the scientific parameters of hazard in the form of probabilistic data to communicate the probable frequency at which a particular district would be exposed to a hazard based on historical data. The level of exposure is marked were classified into three categories Low, Medium, and High based on the damaging intensity and frequency threshold of a hazard. Since the district with the higher damaging threshold and minimum return period are highly exposed they were considered in this study.

Heatwaves: The heatwave trends across 103 weather monitoring stations based on the daily maximum temperature between 1961 and 2010 were mapped by Pai, Srivastava, and Nair.[39] The point data was extrapolated as the district-wise heat wave trends.

Droughts: The district-wise drought trends were taken from the study involving the standardized precipitation index (SPI). This particular study utilized monthly rainfall data from 640 districts of India for 115 years (1901–2015). Districts with a significant decreasing trend in SPI (cumulative) at the significance level of 95% were mapped in the matrix.[40]

In the 78 districts identified from stage one, the associated climate extremes of floods, droughts, and heat waves were identified to have significant impacts on the 10 districts (Fig. 2.2) in the country. The districts exposed to 5 out of all six parameters (i.e. stage 1 and stage 2) were Nagaon, Cachar, Darrang (Assam), Pathanamthitta, Kozhikode (Kerala), Chandel (Manipur), Kanpur Nagar, Bareilly, Kaushambi (Uttar Pradesh), and Uttarkashi (Uttarakhand). These 10 districts were mostly exposed to floods and droughts along with variability in temperature and rainfall.

Stage 3: Extreme Events

The exposure of the 10 districts identified in the previous stage to tropical cyclones and Extreme rainfall events. This stage played a supplementary role in narrowing down the district with higher exposure to multiple climatic events. We considered extreme rainfall events and cyclones as they can be episodic as well as events of higher destruction that involve multiple hazards. We examined the status of exposure of the 10 districts identified in stage 2 to these two events. The criteria involved are as follows:

Extreme Rainfall Events: Up to the year 2012, the data points were obtained from a report on "Extremes of Temperature and Rainfall and the occurrence of extreme rainfall events" (IMD, 2016)[41] by IMD and the information on the events after 2012 was derived from annual climate summary reports published by the department. The extreme rainfall event with the highest magnitude which has ever been recorded at

---

[38] Global Facility for Disaster Reduction and Recovery (GFDRR) (2021).
[39] Pai et al. (2021).
[40] Guhathakurta et al. (2017).
[41] Indian Meteorological Department (IMD) (2016).

2.5 Why Kozhikode?

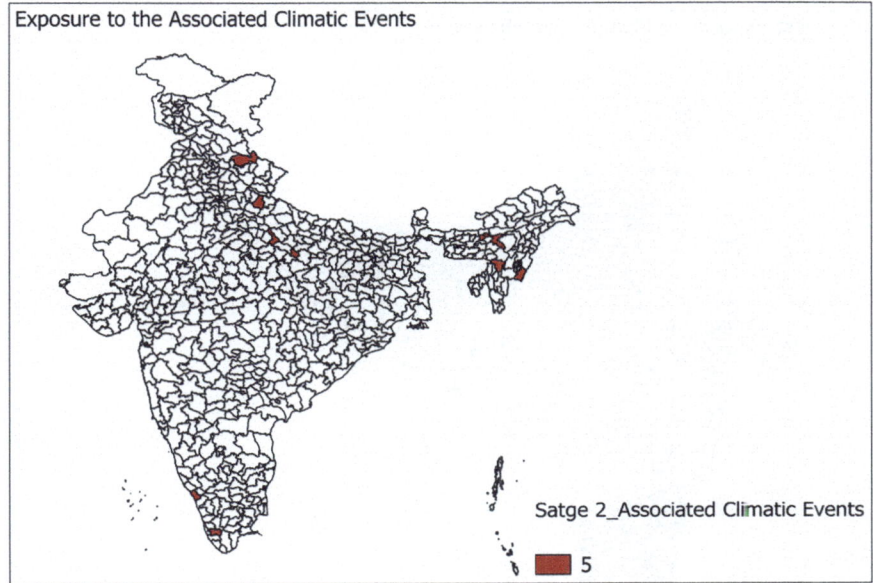

Fig. 2.2 Districts exposed to associated climatic events

a particular station for 24 h located in the 10 districts mentioned above was marked in the matrix.

Cyclones: The trends in the cyclones were obtained from the same source as the floods in stage 2. The districts with the higher probability of the incidence of the hazard with a higher damaging threshold and lower end of the return period were mapped in our study.

The results of this stage enabled us to narrow down to two districts Cachar in Assam and Kozhikode in Kerala (Fig. 2.3) as the most exposed to multiple climatic events in the country. Two of them were found to be exposed to 7 climatic events.

Limitations with data in mapping exposure: The secondary data sources were available for most of the districts in the country. Still, a few of them had to be left out due to the lack of availability of the data. For instance, the district-wise temperature trends for the states of Arunachal Pradesh, Mizoram, and Nagaland could not be found.

2.5 Why Kozhikode?

Among the two districts identified as the most exposed to multiple climatic events. We chose the Kozhikode district in Kerala as the area of interest due to the diversity of its terrain. The district Kozhikode is situated on the southwest coast of India, on

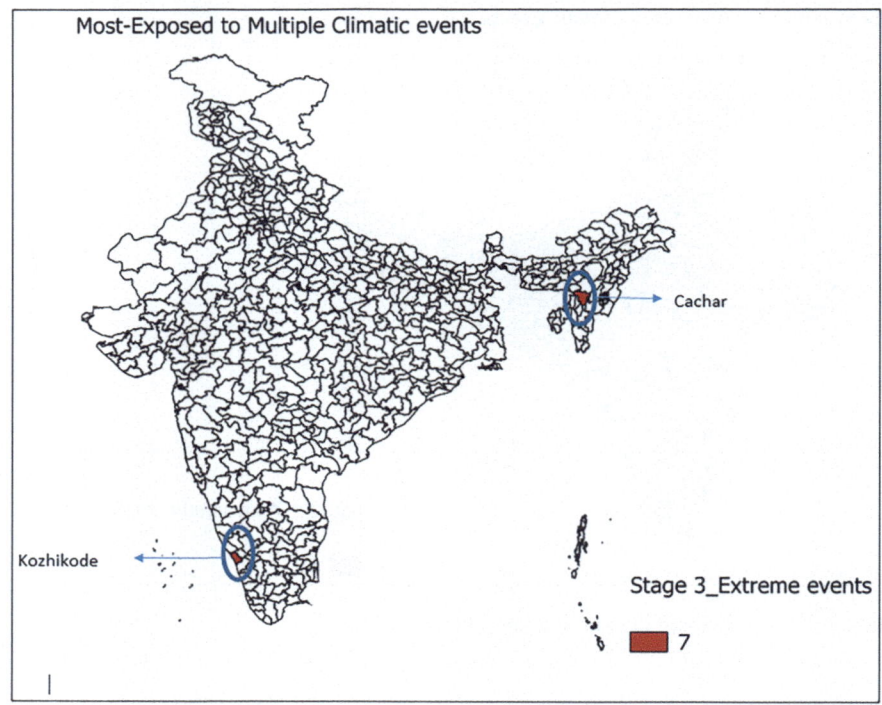

Fig. 2.3 Districts most exposed to multiple climatic events

the northern side of the state of Kerala. It has a long Arabian Sea coast on the west and the Western Ghats on the east. The Kozhikode district shares a border with the district of Malappuram in the south, Kannur in the north, and Wayanad in the east. The total land area of the Kozhikode district is 2344 km^2, of which 362.85 km^2 are sandy coastal lines, which people and communities who depend on the marine ecosystem largely inhabit. Of the remaining land area, a large portion i.e. around 1343.50 km^2 is composed of lateritic midland, which is the area where historically the agrarian villages, emerged and developed centered on paddy cultivation. And 637.65 km^2 areas cover the rocky highland, which is known for the cultivation of spices, plantations, and cash crops.[42]

There is one corporation, four municipalities, and eight gram-panchayats located in the coastal regions of the Kozhikode district. Among them, the Melthura Municipality, located on the north side of the coastal line has one of the longest coastlines. The Melthura municipality has seven wards located on the sea coast. Among these wards, Ward 24 has 380 out of the total 472 households dependent on the Sea for their subsistence, which is relatively highest compared to the other wards in the municipality. Besides, Ward 24 is in proximity to the municipal township of Melthura and

---

[42] Topography|Kozhikode District Website|India.

2.5 Why Kozhikode?

the nearest fishing harbor located in Thikodi, the neighboring gram-panchayat. By considering the population dynamics and the nature of its location, we choose this ward as a potential site in the coastal area.

It is found that the Thamarakulam gram-panchayat, located in the midland area of the district and belonging to the Koyilandy taluk, is a potential site for the study in the midland region of Kozhikode district. The gram-panchayat has a total of 12 different large and small paddy fields, which cover the 500-ha area. 10 out of 15 wards in the gram-panchayats have paddy fields. It has been found that the larger areas of wards 1 and 2 of the Thamarakulam gram-panchayat. Among these two neighboring wards, we choose ward 2 as a potential site by considering its social diversity presents diverse Hindu caste groups and Muslims.

There is one municipality and 15 g-panchayats in the valleys of the Western Ghats in Kozhikode district. Cash crops and plantations are the major cultivation in this area. Hence, had higher proportion of its population is engaged in non-agricultural works, as the criterion for choosing the potential site from the hilly areas. As per the census report of 2011, the Anappara gram-panchayat in Koduvalli block in Kozhikode district has 86.66% of Non-agricultural workers, the highest among the other gram-panchayats located in the hills of Kozhikode district.[43] The Anappara gram-panchayat has a total of 21 wards. Among them, ward 1 has 762 households, which seems highest in the gram-panchayat. Ward 1 is also located in high areas of the gram-panchayats and it has a large number of people engaged with the non-agricultural works.

The major methods used to build on the empirical and theoretical construct of the study were ethnographic fieldwork and household surveys. The stratified random household survey, of 221, 215, and 316 households from Melthura, Thamarakulam, and Anappara respectively, was carried out to obtain the socio-economic dynamics of each village which aided in understanding the level of dominance of different identified situated in the village and thus the dynamics in inequalities, negotiations and conflicts. The survey employed a structured questionnaire obtaining details on the socio-economic status of the households and demographic occupation of the individuals in the households, land holdings, and migrants. The ethnographic fieldwork has been carried out in three months in each village. It employed participant observation, extended long open-ended interviews, and case studies to collect insights on the local history, the process and dynamics of social transformation, and so on in each village.

---

[43] Directorate of Census Operations, Kerala, "District Census Handbook—Village and Town Directory, Kozhikode".

Chapter 3
The Terrain and the Social Structure

As understood in the previous chapter the exposure to climate change is conceptualized upon understanding the impacts of climate change on the people situated at different places as well as the related social-economic systems. The natural climatic processes of a region have been impacted by the latitude where it is situated, the differentiation in the altitude of the landforms within, and proximity to the sea, etc., the impact of climate change varies across the globe while some of the locations are exposed to an increase in rainfall and others face warming and frequent droughts. Likewise, the responses of the terrain to the impacts of climate change also tend to vary. The climate projections have indicated that when the global temperature reaches a tipping point it would be detrimental to the Anthropocene and the ecosystems. The local and regional variability in temperature and precipitation due to global anthropogenic changes in the natural processes of climate has a widespread impact on the people and ecosystems.[1]

A terrain is considered as a lay of land of specific features in the physical geographical sense. Terrains may include anything from mountains, plains, plateaus, etc. But on the other hand in the human geographical sense terrain is the place where the social relations play out and are shaped.[2] This could mean that any changes imposed on a terrain might have effects on the functioning of the social relations and modifications in the social structures as well. Gaining an understanding of the historical and existing social structures and the resultant social relations would aid in understanding existing inequalities and differential responses to the effects of climate change. As each terrain offers a unique set of ecosystem services, the interaction between the populations and their natural environment is determined by the essence of various elements and processes pertinent to a particular terrain. When the discourse about the natural processes impacted by anthropogenic activities has been considered, the

---

[1] Research (UCAR).
[2] Wolch and Dear (2015).

impacts of climate change not only lead to physical changes in the natural environment but also can enable societal impacts.[3] The impacts of climate change are intertwined within the socio-economic processes and when the nature of terrain is diverse, their experience of climate change will also vary.

As we have identified Kozhikode as one of the most exposed to both long-term and short-term climate events, considering the nexus between the nature of the terrain and climate change impacts, the current study engaged with diverse terrains of varied nature. The nature of a terrain could act as a basis for understanding the experience of and responses to change in the climate at the local and regional level as it could play a role in regulating the complexion of the socio-economic processes of a location. The nature of terrain can create differences in the climate across small distances. Within the district of Kozhikode, a coastal village that largely thrived on fishing and related activities, a paddy cultivating village on the plains characterized by a complex wetland ecosystem, and finally, a village that traditionally evolved around the plantations conducive climatic and morphological conditions has been picked up for observation. As topographic diversity is linked with resilience and response to climate change,[4] our first step is to map the diversity of the terrain and the different social structures that historically have emerged in these terrains.

3.1 Melthura: A Fishing Village in Kozhikode

Melthura is a fishing village in Payyoli Municipality, located on the northern part of the coast of the Arabian Sea in the Kozhikode district. Around two decades ago, the village was considered one of the important fishing villages in the district. Melthura had a sizeable stretch of sandy beach. The coastline and the nature of the tide in this part of the Arabian Sea were favourable for the easy sailing of fishing boats. The rock formations near the coast of Melthura attracted many varieties of fish with their feeders and ecosystems to lay their eggs. All these acted as contributing factors in the formation and development of a fishing beach and a fisherfolk settlement which eventually became Melthura.

Historically, the Arabian Sea had a prominent role in shaping the economic life of the majority of people in the village. Along with the fisherfolk from Melthura, the sea-based economy also involved people from the nearby villages in managing businesses related to fishing such as fish selling, selling and repairing of fishing crafts and gears; storage and transportation of the fish; industries related to fish-based products, etc., Apart from fishing, the village and the surrounding area has sandy, soft soil favourable for growing coconuts and cashews but not major food crops. Coconut is the main crop seen in the village, while the cashew trees are scattered and rare. However, the fishing communities in the village engaged less in cultivating these cash crops. The Puslan Muslims and Hindu Mukkuva are the two

---

[3] Krol et al. (2006).

[4] Cai et al. (2018), Lawrence et al. (2021).

fishing communities in Melthura; both have their settlements close to the coastline. The Puslan settlement is on the south side, and the Mukkuva settlement is on the north side of the coastline. Further away from the coastline, after the fishermen's settlement ends, is the area where the non-fishing communities inhabit the village. The Mappila Muslims, Hindu Thiyya, Hindu Viswakarma, and Hindu Pulaya are the non-fishing communities inhabiting the village.

The rock formations in the sea which is nearly 12 kms from the Melthura beach. The local people call it *Velliyamkallu*, which literally means silver rock and has many significances in the fishermen's lives. The Silver Rocks had religious importance since the mythical story behind the Sre Kurumbha Bhagavathi temple the prominent Hindu temple belongs to the Mukkuva community and was associated with the Silver Rocks. A folktale prevails in the village tells that, centuries ago, Bhagavathi, the deity in the temple was halted on the Silver Rocks while travelling from the famous Kodungallur Bhagavathi temple-located in the precent day Thrissur district towards the north side through the sea, and the temple was constructed as the wish of the Bhagavathi. Until the introduction of mechanized boats in the village, in the mid 1980s, the fishermen of the village considered the Silver Rocks limit of their sailing, as these formations hosted varieties of fishes. The fishermen in the village recall that in the off-season they used to go around *Velliyamkallu*, and they were able to catch to survive in the crisis time.

3.1.1 The Social Structure of Melthura

Six communities inhabited the Melthura village. Among them the four communities viz; Thiyya, Mukkuva, Pulaya, and Vishakarma are the Hindu castes, and the two communities—Puslan and Mappila are the castes among the Muslims in North Kerala. Each of these communities, except the Hindu Mukkuva and the Muslim Puslan, historically had distinct occupations and there existed strong social cleavages between these communities. The Mukkuva and the Puslan are the traditional fishing communities and they live close to the coastline in the village. The Mappila Muslims and Thiyya resided close-to-market areas that were largely dependent on non-fishing activities. The other communities—Pulaya and the Viswakarma had a minor presence in the village, as they were relocated to this village in the post-land reform regime. The majority of these communities—Mukkuva, Puslan, Mappila, Thiyya, and Vishwakarma are considered the Other Backword Class and the Pulaya are the Scheduled Castes. However, in the village contexts, the relationship between these communities had specific hierarchical arrangements.

The Mukkuva and the Puslan, though they follow two different religious beliefs, they share many things in everyday life. Irrespective of their religious and political differences they maintained harmonious working relations in the past. Members of these two communities were partners on boats. They sailed together for fishing in the same boat. The *Kadal Kodathi*—literally mean 'Sea Court', is the village council of the fishermen community, a non-state and traditional conflict resolution system in the village, included members from both communities. The council convened regularly on the premises of the sanctum sanctorum of Sri Kurumbha Baghavathi temple in the village. Both Hindus and Muslims attended those meetings. Fishermen had great respect for the members of the *Kadal Kodathi* irrespective of their religion. Another example of the shared co-existence between the Mukkuvan and the Puslan was found in the shared surnames that many of the Hindu and Muslim fishermen still possess in the village, even though they are not living close by. *Koyasankandi* is the name of the small settlement area in the village, which forms as surname of the members of more than five households among both Mukkuvan fishermen and Puslan fishermen in the village.

Both Mukkuvan and the Puslan fishermen communities had followed a joint family system until recently. Under this system, the number of members inhabiting the household was determined by the social and economic status of the unit. The elder member of the joint family—the patriarch played several roles and had varied responsibilities and levels of authority in the family. The joint household owned fishing boats, and the patriarch of the family held the same position in the boats as well. The family members were employed in the boat; their share of the catch was often managed by the patriarch. The family members who were employed in the boat only received money for daily personal expenses. The patriarch has all the responsibility to take care of the day-to-day activities of the family. Many fishermen recalled, that it was the responsibility of the patriarch to manage all the expenses and celebrations in the family which included both life cycle events and calendric festivals.

The Mappila Muslims and the Hindu Thiyya are the two other major communities inhabit in the village. The Mappila Muslims historically controlled the market and trade in and around the Melthura village and the nearby town. A few families of Mappila Muslims also emerged as land elite in Melthura. The Thiyya were engaged in versatile occupations in the village. A few families were engaged in trade, some were traditional healers, and a large section of the Thiyya were engaged in toddy tapping, coconut-related jobs, and traditional house construction using coconut leaf and palm leaf. There were a few affluent families of Thiyya in the village they held a larger portion of the land in Melthura. Two elite families of Mappila Muslims and Thiyya in Melthura also owned fishing boats, which they leased out to the fishermen in the village. These boat owners received a 50% share of each catch. Some affluent Mappilas and Thiyyas even provided financial assistance to the fishermen when they went through financial crises in the off-season. Members of financially weaker families of Mappila Muslims and Thiyya also found employment in the fishing boats.

3.1.2 Ways of Fishing in Melthura

In the early 1980s, when mechanized boats were yet to reach Melthura, the traditional cotton fishnet was just replaced by nylon nets. *Thanduvali* was the major fishing method that profoundly existed in the village at the time. This method of fishing required two units of boats which consisted of an equal number of people varying between 15 and 20 in each boats. Aged fishmen recalled that, there were five elite fishermen families in the village they owned two boats each in the early 1980s. The family members and fishermen from the village were employed in these boats. In addition, six other families were also owned single boats. Those who owned single boat, made partnership with the other single boat owing families in a season basis and some cases they extended the partnership for multiple seasons. There were also two families who specialized in line fishing in the village at that time. The fishermen who specialized in line fishing were aimed at bigger fish such as kingfish, shark, etc.

The technology of fishing was subjected to drastic transformation in the mid-1980s. People recall that the propeller reached the village in 1984, it enabled them to go beyond their usual limits of fishing, and it also helped them to go fishing more than once in a day. Within two years, all the traditional fishing boats in the village had a propeller on them. The introduction of ring seine and *Chundan vallam*-specially designed Plank Canoe, reached Melthura in 1989, which was another revolutionary moment in the life of fishermen in the village. The Government of Kerala's intervention in the Integrated Fisheries Development Project (IFDP)[5] has a pivotal role in bringing this technological transformation to the village. The new fishing technology reached the village in the late 1980s, as a part of the IFDP project which was supported by the National Skill Development Corporation (NSDC), and provided training, and financial aid for newly registered fishing societies to use the advanced fishing techniques. A large size *Chundan Vallam*, ring seine, and two carrier boats were introduced in the village in 1989 by the government-aided fishing societies.

The state-oriented initiatives provided advanced technology to the fishermen, eased their labour, and increased their productivity to improve their social, economic, and health conditions. However, the way the project was implemented in the village provided comparatively more benefits to the non-fishing communities than the fishing communities. Most of the fishing cooperative societies formed in the village at this time had political party affiliations. *Chenthara, Thejwaswini,* and Red Star were the first fishing units introduced in the village and which was mostly controlled by the Centre of Indian Trade Union (CITU), a trade union affiliated to the Communist Party of India-Marxist (CPI-M), a ruling party in the state at the time the project had been introduced. The non-fishing Thiyya community in Melthura was largely allegiant with the CPI-M therefore they were the immediate beneficiaries of the project in Melthura. However, by 1994, all the traditional fishing units moved to using the new technologies. The traditional fishermen acquired the capital for the advanced technologies in two ways. The Muslim fishermen, loyalty to the Indian Union Muslim League (IUML), and the Hindu fishermen's loyalty to the Bharatiya Janata Party (BJP) led to the birth of new fishing societies and they got government aid and support when the Indian National Congress (INC)-led United Democratic Front (UDF) came to power in Kerala in 1991. The old elite, those who owned the traditional boats in the village adapted to this crisis differently. The old elite families-including the Hindu fishermen and Muslim fishermen joint in hand and made new

---

[5] Kurien (1985).

partnership and brought technologically advanced *Chundan Vellam*-trawler and ring seine. *Panathalarajan* and *Thakbheer* were two such boating units introduced in the village by the old elite fishermen families, in the early 1990s without any government aid. While Muslims were the only investors in the *Thakbheer* boat, both Muslims and Hindus were partners in *Panthalarajan* boat.

3.2 Thamarakulam—The Paddy Cultivating Midlands

Thamarakulam is an agrarian village located on the Perambra block, situated on the banks of Kuttiyadi River, and has the largest area of cultivable paddy fields in Kozhikode district. In popular parlance, the village was also called the paddy bowl of North Kerala and the Kuttanad of North Kerala—both indicate the significance of paddy production in the village. The *Kole* land is a natural ecosystem to cultivate paddy, around which the Thamarakulam has been developed as an agrarian village since medieval times. Between the two paddy production seasons, the paddy fields were converted into a vegetable garden, where they produce seasonal vegetables such as vellari-yellow cucumber, vendakka-lady's finger, pumpkin, green chilly, valli payar-yard long bean, tomato etc. The village also has a large area of *Kara* land that was favourable for cultivating coconut, banana, tapioca, and so on.

Historically, the Thamarakulam village has been developed around its paddy field. A composite social world has emerged aligned with the agricultural production in the village. In the feudal regime, the Thamarakulam has many complex forms of land tenure system. The entire paddy field in the village was owned by a patriarch of the Koothali Moopil Nair family, they lived in a village five kilometres away from Thamarakulam. The land was gifted as a *Kanam* to a Namboothiri Brahmin joint family in Thamarakulam. *Kanam* is a specific right over land where the tenant occupies and uses land for a fixed period in return for payments to the landlord. He Namboothiri Brahmin family was not engaged in agricultural production they leased out the entire paddy field to elite Nairs, Nambiar, and Mappila Muslim families—they called *Verumpattakkar*-literally mean only tenant. In return, the Namboothiri Brahmins received a share of the paddy crops. The Namboothiri Brahmins were engaged in managing the temples in the village. The village has the presence of the Nambeeshan community, they assisted the Namboothiri Brahmins in conducting rituals in the temples. The *Verumpattakar* also were not directly engaged in the paddy production, instead, they subleased the land to the cultivating tenants—they were known as *Pattakkar*. The *Pattakkar* were the Nambiars, Nairs, and Mappila Muslims, they cultivated paddy by employing landless agricultural labourers they were largely the Pulayas, landless Thiyyas, and Mappila Muslims. Besides, the village also has the presets of Vaniya—the traditional oil makers, Chaliya—the traditional weavers, and Vishwakarma castes, which include the Thattans-goldsmiths, and Asharis-carpenters.

3.2 Thamarakulam—The Paddy Cultivating Midlands

3.2.1 Social Structure of Thamarakulam

In medieval times, the village primarily thrived on paddy cultivation and the social composition of the village resembled the feudal system. The significance of agriculture, specifically paddy cultivation, has declined over time in the village. Non-agricultural labour become more significant than the agriculture since 1970s.

Eleven communities are inhabiting the Thamarakulam. Among them, nine communities—Namboothiri Brahmins, Nair, Nambiar, Nambisan, Vaniya, Chaliya, Thiyya, Viswakarma, and Pulaya are Hindu castes. The village also has a significant presence of Mappila Muslims. In the last two decades, the village also has the presence of a Latin Catholic Christin household, however, they were not part of the traditional social structure of the village.

3.2.2 The Jenmi System

Before land reform took place in Kerala in the 1970s, the organization of the land relations in Thamarakulam resembled the existing hierarchical land tenure system of the Malabar region. The Namboothiri Brahmins had the highest caste position in the village. The Namboothiri Brahmins didn't have significant numerical strength in Thamarakulam, but they controlled the entire paddy field and other cultivable lands as the *kanakkar* rights over these lands were provided to them by Moopil Nair, to whom the land belonged. The Koothali Moopil Nair was the landlord who owned a major area of the erstwhile Kurumbranadu taluk, of which Thamarakulam village was part. The *Kanam* right over the Thamarakulam paddy field was gifted to the Brahmins, they were not directly engaged in the production of the paddy but collected the portion of paddy from the cultivators as the tax for the land. Nambiars and other Nairs and Mappila Muslims were the cultivating tenants in Thamarakulam. At the bottom of the land relations, some agricultural labourers primarily belong to Pulaya, Thiyya, and landless Muslims. In the exploitative feudal agrarian relation that existed in Thamarakulam, the landless agricultural labourers didn't have any right over the land they worked on or the land they lived in.

3.2.3 Land Reforms

The feudal land relations that existed in Thamarakulam for centuries underwent a significant transformation following the implementation of the Kerala Land Reform Amendment Act in 1970. The oral history revealed that the Koothali Moopil Nair, the patriarch who owns the entire paddy field in the village, and Meloor Illam of the joint family of Namboothiri Brahmins, who were the *Kanankar*, the supervising tenants, had lost their feudal rights over the paddy field in Thamarakulam after the enforcement of the Land Reform Act. Narayanan Nambiar, an 82-year-old cultivator from a Nair community, recalled that Kutichalil Narayanan Nambiar, KC Kunjikrishnan Nayar, the Karanavar—the patriarch of a Nambiar and Nair family, respectively—and Vayayil Abdula Haji and Kunjamad Haji, the Karanavar of the two Mappila Muslim families, were the major beneficiaries of the land reform in the village (Nambiar 2022).[6] These families were the Verumpattakkar—the rich cultivating tenants in the feudal regime. The Nambiars, Nairs, and Mappila Muslims and the Vaniyas became the new landowning elite in the village in the post-Land Reform era.

The landless agricultural labourers from the Mappila Muslims and Thiyya were able to get relative benefits from the land reforms through two processes. These communities in the social hierarchy were close to the cultivating tenants, and that helped them gain some portion of both homestead and cultivable land in the village. These communities had much numerical strength and political dominance, which helped them negotiate with government agencies and landlords. The Pulayas are the least beneficiaries of the land reforms. In Thamarakulam, Pulayas live in four settlements that are similar to what Mencher called the Cherry in Tamil Nadu. These settlements are called Colonies, and each of these Pulaya colonies in the village has 10–25 houses with tiny pieces of land attached to each house. Thus, returns to labour remained the same.

3.2.4 Green Revolution

The paddy cultivation in the village began experiencing the influence of the Green Revolution by the late 1960s. However, a larger-scale shift to the processes of the Green Revolution happened by the mid-1970s after an irrigation canal was opened to the village from the Kittiyadi dam. The Green Revolution increased productivity and improved the agricultural returns of the paddy in the village, which provided more benefits to the new landlords—the Nairs and Muslims.

Besides the increase in productivity, the practice of paddy cultivation has been subjected to drastic transformation during the Green Revolution. Narayanan Nambiar, a 74-year-old cultivator in the villages, recalled that the green revolution had eventually replaced the local varieties of paddy, such as *Vethandan, Odisha, Cheeteni, Thavallakannan*, and so on, with new high-yield varieties such as IR8, *Jaya, Pavizham, Kanchana, Thriveni, Uma,* and so on at different times. The new seeds provided five times more yield than the old varieties of seeds. The cultivators also noticed a significant difference in cropping time. The high-yield paddy seeds only had 120–140 cropping days as compared to the traditional cropping seeds, specifically the *Vethandan*, which was cultivated in the larger area in the village in the pre-green revolution period and has a cropping time of around 11 months. Further, in the old days, the cultivators were the custodians of the paddy seeds; they were

---

[6] Insights from Interviews from field work carried out by Nisar in 2022.

aware of the treatment of the seeds and various cultivation processes, whereas now the state agencies are the new custodians of both seeds and knowledge. More importantly, the cultivation, which was previously heavily reliant only on the monsoon, is now also reliant on canal irrigation.

3.3 Anappara—The Hillslope in Kozhikode

Anappara is a village located on the sloppy hillside of the Western Ghats in the Kozhikode district. The hilly terrain on the Western Ghats in the district was forest land and historically inhabited by indigenous people such as Paniya, Kurichya, and Malamuthan since the time unknown. Until the first half of the twentieth century, any large-scale agricultural production was absent on this part of the Western Ghats in the district. The lands were owned by local feudal elites and they collected the forest products from the indigenous people who lived there. Anappara, as a village in the valleys developed gradually since the 1940s. The terrain is located on the downhill of the Lekidi village on the Wayanad plateau, which used to receive the highest amount of annual rainfall in the State of Kerala. The early settlers of Anappara recalled, that at the time they settled over there, encountered unfamiliar weather events at the location. They experienced six months of intense and continuous rain in the monsoon, three months less rain, and the rest of the months in a year they had to deal with the dry season. They also experienced heavy wind in the month of January, which they called as *January Kattu*.

The slopes of Anappara are characterized by the West Coast Tropical evergreen forest an intermediate between tropical evergreen and moist deciduous forest. However, the predominant nature of the flora found in Anappara is deciduous in nature, as most of the trees become leafless in the dry season. The Anapara stream, which originates from the top of the Anappara hills, flows through the village to a favourable ecosystem to grow abounded flora and fauna in the valley.

The climate in the hilly terrains of the Western Ghats was suitable for growing rubber.[7,8] This attracted the feudal planters towards the terrain in the early 1940s, and they began large-scale rubber plantations on the slopes of Anappara. The larger area of the forest land was converted into rubber plantations in the region at this time. These large rubber plantations belonged to three feudal elites until the early 1970s. Among them, Murikkan, a Christian, and a big cultivator, from Alappuzha which at the time was a part of the Princely State of Travancore, owned a large part of these plantations. Then a Muslim feudatory from Malapuram and a Hindu feudatory from Kozhikode owned the rest of the rubber plantation in the village. The labour employed by Murikkan was largely Christian from south Kerala. Similarly, the majority of the labourers in the plantation that was owned by the Muslim feudatory were Mappila Muslims and the majority of the labourers in the plantation owned by the Hindu

---

[7] Ray et al. (2014).
[8] Sabu et al. (2008).

feudatory were Hindu Thiyyas. These labourers often lived in the housing quarters provided by the owners and these labourers were the early settlers of the village. Anappara market gradually developed on the downhill where churches, schools, and other institutions also emerged along with it.

In the early 1970s, the Anappara witnessed labour conflicts on the plantation. It inevitably forced the feudatories to sell out their land to newly emerging cultivators. Around the same period, the hilly terrain of Western Ghats in the Malabar area also witnessed the second wave of settled migration of Christians from south Kerala. Anappara became one of the places where migrant Christians from south Kerala, Muslims, and Hindus from the neighbouring areas founded their settlements. These settlers bought land from the planters and they further introduced the diverse cash crops and food crops in the village. Since the early 1990s, the Anappara village has been producing areca nut, coconut, pepper, banana, tapioca, cocoa, and coffee at a larger scale in the district.

3.3.1 The Social Structure of Anappara

The establishment of three plantations brought plantation labourers from different parts of Kerala to the village. These labourers were primarily Muslims, Thiyyas, and lower caste Hindus. The second form of immigration was that of the first wave of the Christian migration from the erstwhile Travancore to the hills of Malabar. The hilly terrain of the Malabar area was not explored by the first half of the nineteenth century. The land was available for a cheap price as compared to the land price in the princely states of Kochin and Travancore. This attracted the cultivators, primarily the Christians who were traditionally well-versed in plantation agriculture in the Travancore region and were also facing constraints in the availability of land and other economic crises. The Christians in south Kerala, with the help of the church and kinship network, migrated to the hills across the Malabar region and established new villages there. In the 1940s and 1950s, around twenty families from the Kottayam, Pathanamthitta, and Alapuzha came and settled in the village. These families had average ten-acre land at that time. These early immigrant cultivators and plantation labourers were the first settlers who built the Anappara village.

Anappara was composed of varied social groups. However, all of them are immigrants from different parts of Kerala, they were attracted by either labour opportunities in the plantations or the availability of cheap land. Those who were settled in the village were not economically well off at the place of origin of their migration. These immigrants, with their hard work, had built the village. The neighbourhoods of Anappara characterized by scattered houses enabled less social interaction and the lack of shared common culture between the communities settled in the village, and the differences between the labourers and the planters made the intergroup social relation in the village relatively frail in nature.

3.3.2 The Labour Movement and the Arrival of Small Planters

The decline in the power of the planters provided scope for the growth of the labourers' movement in the plantation in the 1970s. In the 1970s, the big plantation owners had to undergo a severe crisis due to the implementation of land reforms which deteriorated their economic source and power. The bigger rubber planters in Anappara also controlled agricultural land in the plains, which was their primary source of revenue. For example, Joseph Murikan the largest planter, was primarily engaged in paddy cultivation in Kuttanadu. He had 2152 acres for cultivation, similarly, both other planters too had a larger dependency on the agrarian land in the plains. The rise of the labourer movement, and their demand for rights for the homestead land, and better wages had increased the conflict between the labourers and the planters by the mid-1970s in Anappara. The elder people in Anappara recalled that one such struggle turned into a violent conflict in which a labourer was killed by the gangs appointed by a planter. These tensions between the planters and the labourers paved the way for the decline of the larger rubber plantation in the village. The big planters sold their lands to the new in-migrants who came to the village from south Kerala mostly Christians, and nearby areas—Muslims and Hindus. The landless labourers in the plantations also received pieces of homestead land in the Anappara. The fall of the big planters provides a scope for the rice of many small planters in the village. The new small planters largely grow multiple crops depending on the nature of the terrain and soil—it includes coconut, areca nut, pepper, and rubber. Over time they also diversified the crops in response to the market and the changing weather. Cocoa was the major cultivation introduced to the village in the 1990s, and in recent times some cultivators have started cultivating dragon fruit at the place where they once cultivated the pepper.

3.4 Transitions in Social Structure

There have been phases of transitions in power and economic relations among different communities. In this section, we looked at the social structures that existed up until the debate of climate change started to get into the picture globally and when knowledge base on the local effects of climate change had not even begun to build. But when the local effects of climate change started to disrupt or affect the lives of the communities directly or indirectly had to intervene. These interventions were not towards the change in the climate, but largely to the differences in the socio-economic and ecological systems which impacted the livelihoods and survival of the communities. These differences can be at the individual or community level. With the incidence of climate change, the existing social differences and power hierarchies imposed the need for negotiations to formulate necessary interventions. These interventions could further pave the way for a different set of social transformations.

Chapter 4
Sea Change in Melthura

The once-busy fishing beach in Melthura has become a deserted beach now. Creepers and bushes spread across the beach after the fishermen left their harbour due to the difficulties of sailing their boats from the harbour. Now, some old fishermen in Melthura visit the beach in the evening. Some play cards under the shade of the coconut trees; some sit alone, looking at the sea, having deep contemplation, and embracing the wind. There are a few ruined roofless buildings on the beach; with trees growing inside and creepers spread across its wall, where the fishermen and fish merchants once negotiated deals over tea and snacks. The other traces of its old, busy life can still be seen in a few scattered, bush-covered pieces of abandoned boats.

The process which led to this condition of the Melthura harbour was initiated by the changes in the fish catch. The fish catch felt the fluctuations in the availability of notable fish species. The Oil sardines locally known as *Mathi* and Indian mackerels called *Ayila* are the species of interest at Melthura. The qualitative insights from the fishers of Melthura reveal that the fish are not to be found in the locations where they are usually caught. Traditionally the rock formations close to the shore hosted by these fishes do not attract the species during monsoon seasons anymore. The impact and adaptation towards climate change in marine organisms depend on their habitats, geographic affinities, and biological traits. Changes in the species composition of phytoplankton, expansion of the boundaries by small pelagics, species moving to different depths, and phonological changes are some of the biological impacts observed in the Indian seas.[1] Shifts in the geographic range are considered to be one of the most common adaptations to climate change, where it is observed that many species are either shifting or expanding towards higher latitudes.[2]

As a response to the warming of Indian seas, some fish species have been increasing their geographic distribution range. Oil Sardine is known for restricted distribution between 8°–14°N and 75°–77°E which is popularly called the Malabar

---

[1] Vivekanandan (2013).

[2] Adve (2014).

upwelling zone where the annual average SST ranges between 27° and 28 °C. Until 1985 the entire catch of this species was from the Malabar upwelling zone and either few or no catch beyond 14 °N. However, in the last three decades, the catch from 14° to 22°N has been increasing and even accounted for 19% of the all-India catch in 2014.[3] On the other hand, the Indian Mackerel (Rastrelliger kanagurta), a species that is highly sought after next to oil sardines on the western coast of India has been observed to descend to deeper waters along with the expansion of its latitudinal boundary. Between 1985 and 1980 only 2% of mackerel catch was from bottom trawlers,[4] but between 2003 and 2010 this proportion grew up to 15%.[5] This has indicated that the sub-surface warming of the ocean waters led the Indian mackerel to extend its vertical boundary in the last three decades as there in no significant changes in the specifications of gears around that time.[6]

Apart from the movement of the fish species from the changing climate the coastline in Kozhikode district has been subjected to gradual erosion and accretion.[7] The physical changes in the shoreline due to the sea-level rise and the resultant change in the pattern of the tides are observed along the coast of Kerala, 45% of the state's coastline has undergone erosion.[8] The Melthura coast though falls under the medium vulnerable zone to both physical and social impacts of climate change,[9] the ground reality reflects the numerous implications confronted by individuals and communities at different degrees. It is observed that in the last 55 years, the mean sea level has increased by 1.09–1.75 mm/year in the Indian seas.[10] The Arabian Sea region is at a higher risk of Extreme Sea Level (ESL), which is an effect of both an increase in the Mean Sea Level (MSL) and episodic events like storm surges, tides, and cyclones.[11] The ESL exacerbates the risk of coastal flooding and erosion.[12] This is evidence of the regional impacts of climate change and how the global sea-level change eroded the coastline of Melthura and eventually discarded the local harbour.

The gradual increase in the sea level in the last two decades is accompanied by steady sea erosion in the area around Melthura. In 2008, the local government constructed a protection wall at a 3 km distance in Peruthikkad, a village adjacent to Melthura. However, the construction wall could prevent the steady erosion of the beach that was filled with coconut plantations. However, the construction wall had adverse effects on the coastlines in the nearby villages. In a fishing village like Melthura, the consequences were even more drastic. Soon after the construction of the sea protection wall at Peruthikkad, there were sudden changes in the coastline,

---

[3] Hamza et al. (2021); Shyam et al. (2014).
[4] Vivekanandan (2013).
[5] George and Syda (2012).
[6] Punya et al. (2021); Vivekanandan (2010).
[7] Naga Kumar et al. (2022).
[8] Kankara et al. (2018).
[9] Greeshma and Jairaj (2014).
[10] Unnikrishnan et al. (2015).
[11] Sreeraj et al. (2022).
[12] Pugh and Woodworth (2012).

and the tides on the coast became more harsh, especially during the monsoon time. Fishermen in Melthura were not able to adapt to the changes, they were forced to move to a bigger harbour located in Vadakara and Koilandi located 20 and 15 km respectively from the village. Initially, they were moved during the monsoon season, but gradually they became permanent on those two harbours.

Abdullah, a 68-year-old Mappila Muslim fisherman points out, "This village has many advantages, and this stretch of beach in Kozhikode district has the most favourable tides for set sailing. And in this village, we had a larger area of beach. In those days there were no creepers and bushes. The beach was always filled with people, they were engaged in multiple activities. The fishermen coming to the shore after their catch were the kings, they were not engaged in selling it. When the boat reach at the shore, it is the works of *dalal*-the middle men, who fix the price and find market for the catch. Some place they are called *Tharakan*. I remember, there was a long queue of fish carrier trucks, that reached up to the market area located one kilometre away. Now no one coming to this village. We all need to go to other places to catch fish. Now people have started demanding to construct a harbour here. Only then, the old ways can be revived". Out of the total sample survey in Melthura, 10.53% had identified their occupation as fishing as either primary or secondary. Out of them, 28.71% carry out fishing from Melthura itself. A large proportion of the fishermen, 43.56% operate from Chombal harbour and, 26.73% from Koyilandi, and a mere 0.99% from Vadakara.

4.1 Decline of Sardine, Mackerel, and the Loss of Fishing Beach

Oral history revealed that a decline in the catch of sardine and mackerel between 1994 and 1998 created difficulties for the fishermen in Melthura. Sardine and mackerel used to be the primary catch for most of the fishermen depended on for subsistence in the village. Many recalled how a severe decline in sardine and mackerel catch created multiple difficulties in their lives during that time. One of the major effects of this crisis was that most of the government-aided fishing societies failed to repay the loans and some bankrupt. By the end of the 1990s 6 out of 8 boating units were closed as they could not survive the crisis. All these fishing units were initiated by government aid, and the two boating units that survived the crisis were Sri Muruga and Thakbeer, owned by the old elite fishermen families in the village.

But by the mid-2000s new bigger boats which the fishermen refer to as Vanchi, came into the picture. Vanchi can carry forty to sixty fishermen, with inboard engines, and a mechanized system for operating the ring-net was introduced in the village. Technological devices such as cameras to identify and track the movement of fish, GPS systems, and communication devices also came into the picture along with Vanchi. In addition, Vanchi enabled the fishermen to sail up to 40–60 km away into the sea. The Vanchi required high investment ranging from seventy lakh to one crore rupees. Both Sri Muruga and Thakbeer were the only two boating units in the village that were able to buy the *Vanchi*. Both these fishing units were able to find high investment to the upgrade to the Vanchi through the financial investment from the upper-caste elites in the town and financial remittance from the migrants in the gulf respectively.

The long-term change in the shoreline of Melthura paved the way for various difficulties for the fishermen since the latter part of the 2000s. The coast of Kozhikode district serves as the major base for fishing in the state of Kerala.[13] The fishing sector in the district has been facing continuous threats from the impacts of climate change in varied ways in the past few decades. The coastal zones go through complex and dynamic changes naturally over time. However, the climate change impacts and anthropogenic activities have led to unsustainable transformation in these zones.[14,15]

The fishermen of Melthura revealed that traditionally and even now their major catch during the catching season is composed of Sardine and Indian Mackerel. The long-term changes in the physical and chemical properties of the ocean bring about the biological changes in the species that survive within them, as a result, this affects its marine production over time.[16] The wide range of fluctuations in the availability of these species is attributed to the variations in the environmental parameters such as precipitation, upwelling intensity, sea surface temperature (SST), and chlorophyll-a concentration.[17] The Indian mackerel fishery in the Arabian Sea especially around the Kerala coast witnessed annual and decadal fluctuations since 1985, after 1996, the fall was steep while gradually recovering from 2000 with the highest noted landing in 2011.[18] In terms of sardine after a collapse in 1994, faced a major decline in 2013 and 2015. Coupled with the climate change-induced environmental variations, the fish catch at a particular season is also affected by the level and the type of gear involved in the exploitation of the marine resources in the Arabian Sea.[19]

4.1.1 Necessity of Increasing the Scale

The change in the coastline of Melthura had altered the way the fishermen set sail from a particular point on the coast. They also believe that the construction of the coast protection wall in a three kilometre area in a neighbouring village in 2008 by the local government had made sudden and adverse changes in the coastline in their village. With their natural harbour becoming unsuitable for sailing due to changing sea levels. The fishermen sailing with the *Vanchi*-trawlers were forced to move to

---

[13] Salim et al. (2016).
[14] Naga Kumar et al. (2022).
[15] Kankara et al. (2018).
[16] Wabnitz et al. (2018).
[17] Holmes et al. (2021).
[18] Hussain et al. (2021).
[19] Kripa et al. (2018).

the bigger harbour in the first three months of heavy monsoon, and in later months they continued fishing from the village. Fishermen recall that the situation getting worse every year, and people who moved to other bigger harbours gradually stayed depending on those harbours.

People with the smaller fibre boats had been subjected to stall their operations at times as they were not able to sail from their village for most of the fishing season. And, people who were involved in the marketing of fish were also moved to the bigger harbour along with the bigger boats. These factors necessitated the movement towards bigger boats as the only way of adaptation. Fishermen in the village either find employment in the bigger boats or find employment outside of fishing. The fishermen in the village also moved to gulf counties.

Since the 1970s, the village has witnessed labour migration to the Gulf countries. The majority of them were the Mappilas. In the 1990s, a significant number of Puslan too migrated to Gulf countries. Out of the total households surveyed in 2022 from Melthura, 40.27% of them had migrants. The crisis generated by the decline of Sardine and Mackerel seems the pushing factor, many Puslan opt to migrate and find employment there in Gulf countries. However, the representation of Thiyya and Mukkuvan in working in the gulf seems relatively less. The Gulf migration has rearranged the economic structure of the village significantly since the 1990s. It gave birth to new elites in the village the vast majority belonged to the Mappila and Puslan communities. Abdurahiman, who belongs to the Mappila Muslim community was the first person in the village who went to Bahrain in search of livelihood By the early 1990s, the gulf migration had become the significant economic base for the Mappila Muslims who had earlier largely controlled the local trade. The members of Thiyya and Mukkuva also later migrated to gulf countries, their numbers were relatively less as compared to the two Muslim communities. Therefore, there existed growing economic inequalities between the Muslim communities and Hindu communities in the village which become very apparent since the 1990s. The survey revealed that there is inter-group inequality in the way the communities are exposed to the opportunities of gulf migration in the village. Out of the total migrants in our sample, 51.64% belonged to Mappilas and 19.67% to Puslan Muslims. These communities were followed by Mukkuvan (14.75%) and Thiyya (13.93%) respectively.

At the end of the 2000s, all types of fishing activities and related investments were carried on by the villagers themselves. Both Sri Muruga and Thakbeer the two bigger fishing units, operated from the Melthura employed to nearly hundreds of fishermen in their boats. Alongside that, there were also small fishing units operated with small fibre boats. There were more than a dozen small fibre boats in the Melthura village actively engaged in fishing. These fibre boats with the outboard engine also utilized technology such as fish-finder cameras, and GPS systems.

The effects of changes in the coastline had different impacts on people who depend on the bigger boats and people who depend on the small boats. The Sri Muruga and Thakbeer, the two boating units upgraded their Chundan vallam to Vanchi in the mid-2000s, and they both moved to Chombal harbour, a bigger manmade harbour in Vadakara which is located 20 km away from the Melthura village. The upgrading from Chundam Vallam to bigger boats also offered additional labour opportunities.

Both these two boating units provided employment opportunities to nearly hundreds of fishermen in the village. In 2012, when the difficulties of sailing from Melthura escalated, and small fishermen depending on the small boats had no other way to continue fishing activities, three new bigger boats were introduced in Melthura. Among the three new boats introduced in the village in 2012, two were the Marjan and the Ummar owned by a group of gulf migrants, and the remaining one- Shivaganga, largely owned by the Hindu upper caste elites from the Thalassery town.

4.2 Seasonal In-Migration

For the last two decades, that village has witnessed the in-migration of fishermen from Tamil Nadu, particularly during the fishing season. The Latin Catholic fishermen from the Colachel coast of Kanyakumari district Tamil Nadu arrive in Melthura every year in the month of October–November and they base their fishing activities till the end of March. The arrival of the seasonal migrant fisherman in the village is also connected with the changes in the coastline and its other impacts on the village. The changes in the coastline in Melthura had led the local fishermen to abandon the village harbour and migrate to other harbours nearby. The abandoned harbour in the village was attracted the seasonal fishermen to the village.

Since the last two decades, the regular arrival of the seasonal fishermen to the village has had significant impacts on interactions with the local economy- which faced serious challenges after the changes in the coastline and another impact of climate change. The seasonal migrant fishermen specialized in catching *Ayakoora*- King fish, and Squid using techniques and technologies that were not familiar to the local fishermen. Therefore, when local fishermen experienced challenges in fishing due to the multiple impacts of the changing climate, they didn't have many options other than moving to other bigger harbours and getting employment opportunities in bigger boats. The interaction with the migrant fishermen however helped some of the local fishermen to acquire the skills of catching Ayakoora and Squid which a high price in the market. In another way the arrival of the seasonal fishermen to the Melthura also benefited the middleman, many of the middlemen had lost their livelihood once the local fishermen moved to other harbours, now again engaged in marketing fish.

4.3 Proletarianization

The movement towards bigger boats essentially rearranged the labourers in the fishing boats. Unlike in the old times, there was a clear distinction between those who owned the boats and those who provided the labour in the boats. Except for the Sri Muruga and Thakbeer, all other boats, those who invested in the boats are not part of the process of the fishing activities. The fishermen who once had ownership or partnership in small boats become turned into labourers in the bigger boats. This process of large-scale movement towards bigger boats as labourers indeed eroded the autonomy that fishermen once had in the village. This transition has elements of Proletarianization, where the labour class with no access to means of production ends up selling their labour for their survival. One element that makes the process different from a complete form of proletarianization is that the fishermen who provide their labour in the bigger boats still get a share of the catch as they return to labour. It is a contrast to the typical definition of proletarianization, where wage in return for labour seems much more essential. In our survey when the fishermen were asked about how they relate themselves to the boat 19.35% claimed partnership while 80.65% related themselves as labour. In terms of partnership and even labour, the representation of different communities varied. Among the total fishermen claiming to work under partnership 61.11% belonged to the Mukkuvan community while the rest 38.89% was taken up by the Puslans. Even in terms of labour, these two communities held a dominant proportion with 31.67% of mukkuvans and 61.33% of Puslans working as labour in different boating units. Apart from these two the fishing labour also involved Pulayas (2.67%) and Thiyyas (1.33%).

The fishing needs to deal with multiple forms of uncertainties. The catch is not certain, as Vasavan, an active fishermen, once said, "some months there will be a continuous failure, we came back without a single catch, and it will be waste of oil, our efforts, and other investment. But, a good catch in one day some time can provide all of us can recover from the debts." The sharing of catch is a system which the fishermen in the village have been practising from the time unknown. In which, 5% of each catch is a share of God. The capitalisation process however doesn't make any changes to it.

4.4 Consolidation of Identity Capital

The Vanchi required larger investments. Both Hindu and Muslim fishermen had different sources to accumulate capital investment for the bigger boats. The Muslim fishermen were able to attain the capital through the members of their community who migrated to oil-rich gulf countries. Al Ameen, Umar Mukthar, Kaleefa, and Madinah are the four *Vanchis* introduced by the Muslim fishermen in the village. Major investment in these boats came from Gulf remittances. The Hindu Mukuva fishermen were not able to tap the Gulf remittance since there were only a few among them who migrated to Gulf countries. However, two new *Vanchi* have been introduced by the Hindu fishermen in the village. Their capital investment came from the upper-caste Hindu elite in the Thalassery and Vadakara towns. The owners of these boats revealed that they get immense help from the Rashtra Swayam Sevak Sangam (RSS) to get in touch with these investors.

The consolidation of the identity capital- the gulf remittance on the Muslim fishermen's boats, and the Hindu upper caste investment on the Hindu fishermen's boats also reflected changes in the secular social and economic fabric of the village. The source of consolidation of identity capital investment has led to the exacerbation of communal polarization on the coast. Earlier, the Hindu fishermen and Muslim fishermen used to sail for fishing in the same boats, the elite among them even had partnerships on boats, and the Mappila Muslims, the important trade community in the village were the source of immediate financial needs of the fishermen across the religion. But, with the emergence of the identity capital, the village witnessed the initiation of the process of religious polarization. The Hindus and Muslims are no longer going fishing in the same boats. There are no existing partnerships between elite Hindu fishermen and Muslim fishermen currently active in the village. It even reflects in the friendship ties, and neighbourhood interactions between the two fishing communities in Melthura.

4.5 Transformation of the Social Structure

The rearrangements of capital and labour led to subsequent changes in the social structure of the village in the last couple of decades. At present there is no partnership exists between the Mukkuvan and Puslan on owing boats. In 1994, the two Puslan old elite families became partners, with the Mukkuvan in Sri Muruga to overcome the challenges generated due to the introduction Government aided fishing cooperatives largely by the non-fishing Thiyya community with advanced Chundam Vallam and Ring-net. But with the growing religious polarization, Sri Muruga is now considered a Hindu boat, which has capital investment from the Hindu upper caste elite, mostly RSS supporters. The new capital investment from the RSS-supported upper caste elite damaged the earlier economic ties between the old elites of Mukkuva and Puslan After being a silent partner for 4 years, the Puslan finally withdrew their investment from the Sri Muruga in 2022.

Now politico-religious identity become an integral nature of the capital investments on the boats. The identity capital on boats primarily reflected the identity of the labourers in the boats. At present, among the 60 labourers in the Sri Muruga boats, there is no single member from Puslan as a labourer. It is said by Narayanan, a Mukkuvan and a partner of the Sri Muruga that among the 60 labourers 59 are active supporters of RSS and Bhartiya Janatha Party, the one labour belonged to Thiyya who is a supporter of CPI-M. Similarly, among the three boats owned by the group of new elites from the Puslan and the Mappila, in which nearly 150 labourers are engaged, there is no single labour from the Mukkuvan community.

The prevalence of the influence of different identity groups is observed in Melthura. To further dissect the dominance of different identities in the village we looked into the sample surveyed. The range of dominance of various identity groups has been established by measuring the Degree of Identity Dominance (DID).[20] The

---

[20] Pani (2022). The Degree of Identity Dominance (DID) involves average of the indicators of the three instruments such as the Economic power of identity group, Political dominance of identity group and dominance of an identity group in the realm of knowledge.

4.5 Transformation of the Social Structure

results revealed that on the whole, Melthura is found to have been constituted by three prominent communities such as Mappilas (0.41), Puslans (0.23), and Mukkuvans (0.20). Even their numerical dominance follows the same order, with the individuals of Mappila Muslims constituting around 40.35% of the sample surveyed followed by Puslan Muslims (23.88%) and Mukkuvans (19.08%). In terms of economic dominance as well the Mappila households from our sample held 45.18% of the assets, while the Mukkuvan households stood behind them with 20.85% of the assets and the Puslans who were preceding Mukkuvans in numerical dominance were found to be a little behind of the Hindu fishing community with 20.47% of the assets.

Though being identified as a fishing village the residents of Melthura have diversified their occupation based on the changing scheme of things. The occupational structure of the Melthura obtained by analysing our sample reveals that a large proportion of the most prominent group of Mappilas was either employed as Skilled workers (28.13%) or Other wage labour (28.13%), as they are also involved in investing in the fishing boats nearly 16.67% of them identified themselves as own-account workers. 40% of the Mukkuvans were employed in the fishing sector while around 20% of them were employed at different places as skilled workers. In terms of the Puslans, 50.52% of them had fishing as their primary occupation apart from it they had an occupation as skilled workers (15.46%) or own account workers (7.22%). Apart from these three communities, the Thiyya is the other community which co-existed in Melthura. They formed nearly 13.87% of our sample and primarily worked as other wage labour (32.73%) and skilled workers (20%).

Besides the communal polarization, the movement towards the bigger harbour and bigger boats also had other implications across age and gender. The movement indeed excluded the elder people left jobless in the village. The abandonment of the natural harbour also forced the economic activities that women engaged in the old harbour. The loss of these economic activities and the subsequent crisis that fishermen experienced due to the other long-term impact of climate change inevitably forced the women to move out of the village and find employment outside. From the household survey, the representation of women of working age in different types of occupations could be understood. With zero involvement in fishing and agricultural labour, nearly equal per cent of them were involved in their account work (43.24%) and private salaried jobs (42.11%) as compared to men in the working age.

The disconnection of economic activity in terms of the sharing of capital and labour between the Mukkuva and Puslan has implications for the social structure of the village as well. The two communities who once shared neighbourhoods now deliberately moved to both sides of the village. In a short period, the settlement pattern of the village has been subjected to a drastic transformation. The south side of the Melthura became Puslan dominating regain and the north side of the village became the Mukkuva dominating place. This communal polarization indeed generated hatred and animosity among both communities.

Chapter 5
The Shifting Rains of Thamarakulam

In the month of Edavam,[1] the rain comes with thunder and lightening

In Midhunam, the rain will not disappear

The rain brings destituteness in the month of Karkkidakam

The drizzles make the sounds of anklets in the month of Chingam

There will be random rains in the month of Kanni

In the month of Thulam, the rain pours water after measuring it

The rain is chilly when it comes in the Month of Vrishikam

In the month of Meenam, it is Makara Mazha

In the Medam, there will be new rains

The folk song above was sung by Gangadaran Nambiar, a retired school teacher and a part-time cultivator in Thamarakulam, to explain how predictable the village's rainfall was each month until recently. Gangatharan Nambiar lives in a recently modified old ancestral house near the paddy field. Gangadaran Nambiar also has a nearly 10-acre paddy field close to his house, a large portion of the land he got as hereditary, and the other portions he brought later. However, the entire 10-acre paddy field has now been abandoned from cultivation. Gangadaran has experienced multiple difficulties in paddy cultivation for the last one and a half decades. It all started with the flooding at the paddy field, which generated difficulties in paddy cultivation at a larger land area that Gangadaran Nambiar owns. Until 2020, he had cultivated three acres but abandoned them due to the frequent crop failures due to the unexpected rain.

---

[1] Edavam, Midhunam, Karkidakam, Chingam, Kanni, Thulam, Vrishikam, Meenam and Madam are Malayalam months.

The folk song was Gandaran's answer to how the rainfall differs from that of the past. He had explained that the rain was more or less predictable. Cultivation and harvesting of all crops depended on the predictability of rainfall. Gandaran said, *"Nowadays, no one can predict when it will rain and when it ends; when it rains, there will be too much rain, and the very next day, there will be days with bright sun. This kind of rainfall will destroy all the vegetables, not just paddy."* One of the main reasons that Gangadaran stopped his paddy cultivation in the remaining three acres was the unusual rainfall that has frequently damaged crops since 2017. Many cultivators in the village had similar experiences of unusual rain in the form of prolonged monsoon and early monsoon, which generated difficulties in paddy production.

5.1 Changing Rice Fields

The variations in the monsoon- the prolonged monsoon and early monsoon created multiple constraints, especially for the paddy cultivators increasingly and the village has gone through diverse land use changes. In addition, drought, water scarcity, and other transitions in the land use of Thamarakulam forced some of the cultivators to stop cultivation gradually since the early 2000s.

The long-term trends (1901–2015) in both the maximum and the minimum temperature had shown a significant increasing trend in the district of Kozhikode.[2] The increasing temperature has diverse impacts on rice production.[3] The rise in temperature in the early summer is one such issue that the cultivators are grappling with to find solutions. In early February 2023, a significant portion of the paddy field, belonging to the cultivator's society A, was damaged due to the increasing temperature and the water shortage. Moidu Haji, the president of the society, stated, *"Last year it was early rain that destroyed the crops; this year it is the early summer that damaged the paddy; this is the time the paddy needs water in the field, but look at the straw; it all turned yellow because of the summer. I don't think we will get any yields from this".* He later plucked a yield from the paddy saplings, pressed it, and said, *"It is just a paddy cover, nothing formed inside it."*

Vagaries of monsoon is one of the pressing issues faced by farmers in low-elevation areas. The erratic rainfall events cause water logging during the harvest or growing season leading to loss of crop or straw. The rainfall variability and seasonality affect the production of the crops; they are also major determinants of soil moisture.[4] In response to the environmental variability brought on by climate stress, farmers are often forced to modify their sowing and harvesting schedules, seed types, crop rotation schedules, and irrigation techniques.[5] The effective growth and management

---

[2] Indian Meteorological Department.
[3] Saud et al. (2022).
[4] Bedane et al. (2022).
[5] Marcinkowski and Piniewski (2018).

of crops depend on the long-term changes in temperature and precipitation.[6] The late or the early arrival of monsoon impacted the sowing and the harvesting cycles of the crop. The inter-annual variability in the rainfall is observed to be less when compared to the seasonal variability and there is a prominent increase in the rainfall during the summer or pre-monsoon in the district.[7] Which also impacted the overall production and productivity of the crop. In addition, the differences in on-set and withdrawal of the monsoon disoriented the sowing and harvesting timings of the farmers, affecting their yield and leading to crop damage. Similarly, farmers in the medium-elevated areas dealt with postponement of cultivation due to prolonged monsoon, while the early arrival of monsoons affected the harvest. Simultaneously, the people who cultivate in the high-elevated areas are also affected by the drought, and the lack of water in the field affects the growth and productivity of the yield. The rainfall patterns of the district are seasonal and are accompanied by long, drier seasons. According to the Reconnaissance Drought Index (RDI), Kozhikode district was under drought for 9 years out of 35 (1980–2015).[8] The highlands in the Malabar region have become vulnerable to drought in recent years due to climate variability.[9] The long-term variability in rainfall has been observed in the district of Kozhikode to have led to a meteorological drought.[10] Since the 1990s and from 2000 to 2019, the district has witnessed drought of varied degrees.[11] The variations in the rainfall had impacted the production and productivity of crops such as paddy and Coconut in Kozhikode.[12]

The farmers on the lower elevation also had to deal with the saltwater intrusion during high tides. The introduction of saltwater into the paddy field during the days of high tide is a natural phenomenon of the wetlands but the disturbance in the crop cycle presented a new difficulty for the cultivators with the intrusion of saltwater into the fields. There are water level fluctuations in the local aquifers between pre-monsoon and post-monsoon periods in the region. There are water level fluctuations in the local aquifers between pre-monsoon and post-monsoon periods in the region. High salinity levels result from the reverse hydraulic gradient and saltwater intrusion during the pre-monsoon months because the aquifers in the Vadakara-Quilandy stretch, the larger region to which the Thamarakulam geographically belongs, tend to be the evaporation dominant zones.[13]

---

[6] Bannayan et al. (2011).
[7] Surendran et al. (2019).
[8] Gopinath et al. (2020).
[9] Gopinath et al. (2015).
[10] Surendran et al. (2019).
[11] Gopinath et al. (2020).
[12] Abhinav et al. (2018).
[13] Sheeja et al. (2022).

5.2 The Floods of Artificial Recharge

Thamarakulam being a paddy-cultivating village witnessed the effects of rainfall variability in complex ways. The drought resulting from long-term variability in rainfall has affected the water availability in Thamarakulam. With the decrease in surface water sources, the authorities turned their focus on the groundwater. But the strategies devised to deal with the drinking water scarcity in particular proved decisive to the paddy cultivators in the village. The policy decisions to create infrastructure to manage the water resources in the region added problems for the cultivators. In 2004, the State Water Authority constructed a drinking water pump house to extract the groundwater on the banks of the paddy field to deal with the drinking water crisis in the summer. When the demand for drinking water escalated further, two other agencies—the Thamarakulam Gram Panchayat and the Jalanidhi Project—constructed four more such pump houses on the banks of the paddy field in the years 2006, 2010, 2014, and 2018. Currently, two more new pump houses funded by the Jalanidhi project are under construction. Each of these pump houses has wells on the banks of the paddy field and could distribute 50,000 L of water per day.

The continuous extraction of the groundwater led to a significant fall in the groundwater level in the district in recent years. The Kozhikode district on average witnessed a 1–2 m decline in groundwater level in 2021 when compared to the average water level of the previous decade.[14] This resulted in a decline in the water supply from the pump houses. As a result, the water resource management authorities in Thamarakulam were forced to take measures to artificially recharge the groundwater aquifers. The irrigated paddy fields contribute to groundwater recharge in various parts of the country, especially the proportion of seepage from the flooded paddy fields contributes to maintaining the water levels of the groundwater aquifers under them.[15] So, to ensure the undisrupted drinking water supply during summer the paddy fields of Thamarakulam were flooded by releasing water available in the irrigation canal to enable the groundwater recharge. The impact of this execution on water supply highly impacted the paddy fields at the lower elevation and mid-elevation. The stagnated water in the paddy fields created a conducive environment for the spread of algae.

5.3 Responses

One of the responses against the disturbance of the variations in rainfall in paddy cultivation was the return to the old seeds. Some paddy cultivators, who experienced frequent failure of cropping damage due to the effects of the prolonged monsoon and early monsoon, found an old paddy seed, *Vethandan*, which was one of the significant paddy seeds that existed in the pre-green revolution time in the village, a useful to

---

[14] Groundwater Department (2021).
[15] Gulati et al. (2022).

continue the paddy cultivation. The *Vethandan* seeds have a long cropping duration of 11 months, fewer input requirements, large straw output, and high resilience towards floods, which attracted the cultivators, especially those with cattle at home, to opt for this option. The lower labour inputs and high cropping time also benefited those who continued paddy cultivation by opting for this method. However, the *Vethandan* was only a viable option for some of the village's cultivators, who largely depended on the market. Those who depend on the market have many difficulties continuing paddy cultivation.

In the last two decades, a larger area of Thamarakulam paddy fields has been converted into plantain fields. The continued failure of paddy cultivation due to the variation in rainfall is one reason many traditional paddy cultivators moved to plantain cultivation. Plantain cultivation in the villages seems less affected by rainfall variations than paddy cultivation. The market price and the lower labour inputs also accelerated the diversification of the crops from Paddy to plantain in the village, especially in the last two decades. However, plantain cultivation was only possible in the areas that were not affected by the artificial flooding in the paddy field and the salination, and those areas remain not cultivable to plantain or any other vegetable.

The area of the Paddy field affected by the salination is almost abounded by the cultivators. Abdullah, a cultivator who owned 5 acres of a paddy field close to the Kuttiyadi River, had abandoned the entire area due to the difficulties of salination of the field. However, the area affected by the artificial flooding, though the larger areas remain abandoned, has already been converted for fish culture. Three people began fish culture in the flooded field in 2020, growing fish varieties such as Tilapia, Katla, and Catfish.

5.4 The New Social Relations of Paddy Production

The direct effect of climate change on agriculture is in the production of major crops. The combination of unseasonal rainfall flooding and declining water tables has been known to reduce the yield of major crops across the world. In, Kozhikode even though the long-term yield of the rice looks marginally stable the area and the production of the crop recorded a significant decreasing trend. The graph below shows the changes in area, yield, and production of rice in Kozhikode between 1998 and 2020.[16] Table 5.1, showcases the significant decline in the number of cultivators registered under the agricultural societies, as well as the land cultivated in Thamarakulam in the year 2022.

The difficulty in producing the crops brought about by climate change has the potential to fundamentally alter agrarian relations. The decline in area under production impacts cultivators and agricultural labourers very differently, because of the vast divergence in the socioeconomic context in which they operate. The cultivators belong to landowning communities, whether they are the traditional landowners like

---

[16] DES|Area, Production and Yield—Reports.

Table 5.1 The paddy cultivation in Thamarakulam in the year 2022

Name of the cultivating societies	Number of cultivators registered	Total cultivable land (in acres)	Number of people cultivated in 2022	The total area cultivated in 2022 (in acres)
A	190	220	21	30
B	125	150	12	15
C	130	180	40	60
D	196	220	84	110
E	112	112	48	60
Total	753	882	205	275

the Nairs or those who got land in land reforms like the Thiyyas. The Muslims in the village belong to both categories. In contrast, the agricultural labourers typically belong to a scheduled caste, the Pulayas, who have been kept out of any rights to the land. Even when land reforms were carried out, they were, in practice, primarily land to the tenant movement. Not being tenants, the agricultural labourers were largely bypassed in this reform.

Over the past decade, land relations have been subjected to drastic changes in Thamarakulam. As noted earlier, the implementation of the Land Reform in the 1970s abolished the landlord status of the Koothali Moopil Nairs and the Kanakkar rights of the Namboothiri Brahmin joint family which led to the Nairs and Mappila Muslims to emerge as the new land elites in the village. The Nairs and the Mappilas who were either the tenants or the cultivating tenants in the paddy field, now became the land-owning cultivators in Thamarakkulam. Whereas, the Pulaya, Thiyya, and some landless Mappilas remained as the labourers in the field. The relationship between the new land elites and the labourers, however, was not completely modernized until the beginning of the new millennium. Aboobacker, a Mappila cultivator said *"Even at the beginning of the 2000s, the labourers used to come to us and claimed that the work is their family right, that have been working in my field for generations. In those times, money was not given to the labourers as their wage, they came and did their work and took the share of the crops as a return for their labour."*

The Nair and Mappila cultivators revealed that the frequent failure of crops due to various climate change impacts such as the late monsoon, early monsoon, and the perpetuated flood, experienced in the early 2000s led to lower returns due to the crop damage and the labourers involved in non-farm wage labour started demanding similar wage arrangements from agricultural labour. The climate change distress not only changed the traditional feudal reward for labour in the paddy field but also led to changes in land relations and traditional caste-based social composition aligned with the paddy production in the village. The distress in the paddy cultivation forced the landowning cultivators to either abandon the paddy field, migrate to other villages for cultivation, or change the cultivation from paddy to arecanut or banana. Some of them even entered fish culture in their flooded paddy fields. As a result, the land value of the paddy fields declined, and this encouraged the traditional landless agricultural labourers to lease in these paddy fields and become cultivators.

To further present the current demographic and economic conditions of the village, the data from the household survey was analysed. The Degree of Index Dominance[17] revealed that overall the Mappila Muslims fell under the prominent category community in the village with a score of 0.4 while other communities such as the Nairs (0.18), Pulayas (0.16), and Thiyyas (0.13) The numerical dominance of the Mappila Muslims was also observed with them comprising 42% of the sample population. In terms of numbers, the mappilas were followed by Pulaya (19%), Nair (16%) and Thiyya (12%). In terms sum of the assets the Mappila households held 45.05% and Nairs 20.44% followed by Thiyyas (14.47%) and Pulayas (11.05%).

The survey data also revealed the transformation in land relations and occupational structures. Out of the total operational landholdings held by the sample population, 52% were owned by the Mappilas and 22.1% by the Nairs. The traditionally landless communities such as the Pulayas and Thiyyas were observed to hold 5.9 and 9.5%, respectively. We observed that a large proportion (56.3%) of the people from our sample who were cultivators belonged to Mappilas. The Pulayas who have started acquiring lands under the newfound opportunities created by climate change distress and related changes also formed a significant proportion (25%) of the total cultivators in Thamarakulam.

The movement of people who were previously cultivators or agricultural labourers to non-farm jobs did not completely reduce the prospects of agriculture in the village, especially as a source of income. Though the number of people depending primarily on agriculture declined, the majority still see agriculture at least as a secondary occupation in the village. The open-ended interviews conducted with the farmers of Thamarakulam revealed issues such as drought, and significant variability in the monsoon rainfall led to a series of other events like water stagnation, the spread of invasive algae, and saltwater intrusion which made paddy cultivation extremely difficult. We further investigate how the agrarian system under distress enabled a social transformation in the village.

5.5 Beyond the Rice Fields

The reproduction of the social relations and social structure of Thamarakulam has revolved around the paddy field. Throughout the period of the feudal regime, social relations were strongly established hierarchically. That provided benefits to the caste and communities positioned at the top of the social hierarchy and those who provided labour in the paddy field were treated as just goods. The head of the Moopil Nair

---

[17] Pani and National Institute of Advanced Studies (Bangalore, India) (2022). The Degree of Identity Dominance (DID) involves average of the indicators of the three instruments such as the Economic power of identity group, Political dominance of identity group and dominance of an identity group in the realm of knowledge.

family held the birthright to the entire paddy field and received a share of crops as rent. The Namboothiri Brahmin had the highest position in the social hierarchy of the village; the Moopil Nair family was provided the right to receive the share of crops from the Namboorthiri Brahmin family in the village. In addition, there were also elite Nair and Muslim families that had the rights to supervise tenants and also received a considerable portion of the share of the crops. These arrangements inevitably put much pressure on the cultivating tenets—the families that belong to Nairs and Muslims but are not as privileged and powerful as the supervising tenets. This multi-tiered land tenure system has resulted in the creation of a class of labourers who do not have any rights over what they produce. Indeed, the labourers in the field were primarily Pulayas and the members of poor Thiyya and Muslim families in the village. These labourers, in particular the Pulaya, were considered untouchables; they were not allowed to enter temples or public areas and lived on plots of land that their owners had given them on a temporary basis.

This exploitative multi-tiered land tenancy system has triggered a rebellion in some parts of North Kerala since the 1830s, specifically in the area where Mappila Muslims were the tenants and agricultural labourers. In the early part of the nineteenth century, the political movement in North Kerala formed largely on a nationalist ideological base, began to address the social issues of exploitation entrenched in the feudal-Agrarian relationship, since communist ideology gained prominence in the nationalist movement in Malabar. The Malabar Tenancy Act of 1929,[18] by the British Raj, was seen as the culmination of growing tensions in agrarian relations. However, the legislation of 1929 and its amendment in 1951 and the later legislation addressed issues between the landlords and the tenants. The exploitation of labour seems largely ignored in these interventions that aim to bring democratic reforms to the existing exploitative feudal land and agrarian relations. It is a criticism of the Kerala Land Reform (Amendment) Act of 1969[19] that the beneficiaries of it were largely the tenants; the agricultural labourers received only a small piece of homestead land.

Thamarakulam was not an exception to the trajectory of agrarian changes in Malabar. The land reform abolished the rights of landlords and their role in the social relations of paddy production was eroded or completely disappeared. The Nairs and Mappila Muslims, who were the supervising tenants or the cultivating tenants, were the major beneficiaries of the land reforms. The Thiyya and the poor Muslims also received relative benefits from the land reforms. However, the Pulaya, a Scheduled Caste in the village, historically provided labour in the Thamarakulam paddy field and were at the receiving end of land reforms. The difference between the cultivators and the agricultural labourers extended well beyond the rice fields. The divisions were heightened in social practices. The Pulaya, the untouchable caste in the feudal regime, were employed as labourers in the paddy field and were at the receiving end of the land reform in Kerala. The Pulaya people benefited from an average of 5–10 cents of homestead land but not cultivable land. As a result, they remained labourers in the paddy fields in Kerala during the post-land reform period. The role and position

---

[18] Vattarambath (2007).

[19] Ambli (2021).

of the Pulaya in the agrarian relationship of the village were not subjected to any significant transformation in the post-land reform period. The only exception to this was the economic mobility that a small number of Pulaya community members achieved through government employment and education. However, until the early 2000s, Pulaya men and women were the major labour force employed in the paddy field in Thamarakulam.

In the last two decades, drastic transformations have taken place in the village's social structure and the agrarian social relations associated with paddy cultivation. Among many reasons, such as the decline of agriculture in the economic structure of the village caused by the growth of the service sector and the tertiary sector, and the new economic opportunities brought by Gulf migration and its remittances to the village, the changes in the climate and its impacts on agriculture contributed to the structural transformation in the social relations in agricultural production in Tharamarakulam. The decline in the area of production due to the perpetuated flooding, and uncertainties of crop failure due to the rainfall variations led the traditional cultivators—the Nairs and landowning Mappila Muslims, to move out of agriculture or diversify the crops. The movement out of agriculture by the traditional cultivators offered opportunities to the traditional agricultural labourers and also traditional artisans to enter the cultivation as new tenants. Our survey revealed that out of the total respondents in the working-age population who had identified cultivation as their primary occupation only 6.3% belonged to the Nair community. The survey revealed that traditional agricultural labour groups, the Pulaya, landless Thiyya, Mappila Muslims, and Viswakarma have emerged as the major cultivators in the village at present, 56.3% of the cultivators belonged to Mappila Muslims and 25% to the Pulayas.

The qualitative inquiry reveals that the uncertainties of risk have significantly contributed to many traditional cultivators leaving the field. *"Paddy cultivation is not profitable anymore; I didn't even get the coolie I paid for the labours"* is a common feeling among the land-owing cultivators who left the paddy field. This led many of the landowners to lease out their land. The lease amount, too, varies according to the productivity of the field and the resistance to the impact of climate change. The flooded paddy fields are leased out with no expectation in return, and Narayanan and three others have started fish culture. The general rule of tenancy followed in other areas is to share half of the crops. It is also significant to observe that the new cultivating tenants do not intend to do the cultivation as their only engagement. Rather, many of the new cultivating tenants- including Pulaya, Thiyya, Mappila Muslims, and Vishwakarma, are considering it the only economic source. Rather, the majority of them have other means of livelihood as well.

The economic and social marginalization of the Pulayas was predictably reflected in their educational levels. The emphasis that state governments going back to the 1950s laid on education had only limited effect on this community. At the time of the survey conducted for this study in 2023, the average number of years of formal education of the Pulayas was years compared to years for the Nairs as well as the Thiyya. It was observed that Nairs and Thiyyas had better education levels than Pulayas. The differences in education levels had a direct impact on the opportunities available to the different communities from education and migration led to growth in Kerala's economy. In our survey, the migrant population belonged to Mappila Muslims (65.57%), Nairs (13.11%), Pulayas (13.11%), Thiyya (4.92%), and Vishwakarma (3.28%). A stark difference in the average years of formal education between the migrants and non-migrants was witnessed in the results of our survey the average years of formal education of the non-migrant Mappila Muslim was 8.82 the migrants from the same community had 11.93 average years of formal education. Similarly, Pulayas who are non-migrants had 9.59 average years of education while it was 13.25 among the migrants.

To the extent that the discriminatory social practices were grounded in agrarian relations, the Pulayas had reason to prioritize a change in their relationship with the land. At the bottom of the traditional agrarian hierarchy, the Pulayas saw acts of discrimination rooted in their role in agricultural activities. Unlike the tenants who benefitted from land reforms, the Pulayas as agricultural labour were largely unaffected by that major state initiative. The opportunity to become a cultivator rather than just remain an agricultural labourer thus had a social dividend for the Pulayas. The social dividend opened up opportunities for the Pulayas that are best captured in terms of textbook microeconomics.

The differences in economic considerations across social groups in Cheruvannur can be seen in Fig. 5.2. We can begin with the considerations facing the cultivators that is the Nairs, Thiyyas, and Muslims, before climate change. Without the effects of climate change, the cultivators faced a Marginal productivity curve of Y1 Y1'. They would then cultivate up to the point, E1, where the Marginal Cost curve, C1, intersected the Marginal Productivity Curve Y1 Y1'. The coming of climate change contributed to a downward shift in the Marginal Productivity Curve. If the same Marginal Cost Curve, C1, continues agricultural operations are no longer viable.

The conditions facing the Pulayas are, however, different. As we have noted they had reason to expect a social dividend if they were to become cultivators. A relationship with the land as tenants would give them a status in the agrarian hierarchy that had not had before. As cultivators, they could well impute lower wages to themselves in exchange for their changed socioeconomic relationship with land. The lower imputed wage costs would shift the Marginal Cost Curve downwards for agricultural labourers turned cultivators. This is depicted in Fig. 5.1 as C2. At this lower cost curve, agriculture returns to being viable even at the lower productivity after climate change.

The very different situations facing the earlier cultivators and their agricultural labour make a strong case for tenancy. It suits the agricultural labourer to seek to become cultivators by leasing the land. At the same time, it suits the cultivators, who no longer find agriculture viable, to lease out their land. Results from our survey reveal that out of the total land that was leased out in the village the Nairs held a larger proportion (44%) followed by Thiyya (29.41%), Mappila (18.07%), and Vaniya (8.40%), respectively. On the other hand in terms of the lands leased-in the Mappilas stood first by holding 32.04% of the total leased-in land. They were followed by the Pulayas (20.87%), Thiyyas (20.63%) and Nairs (17.64%).

5.5 Beyond the Rice Fields

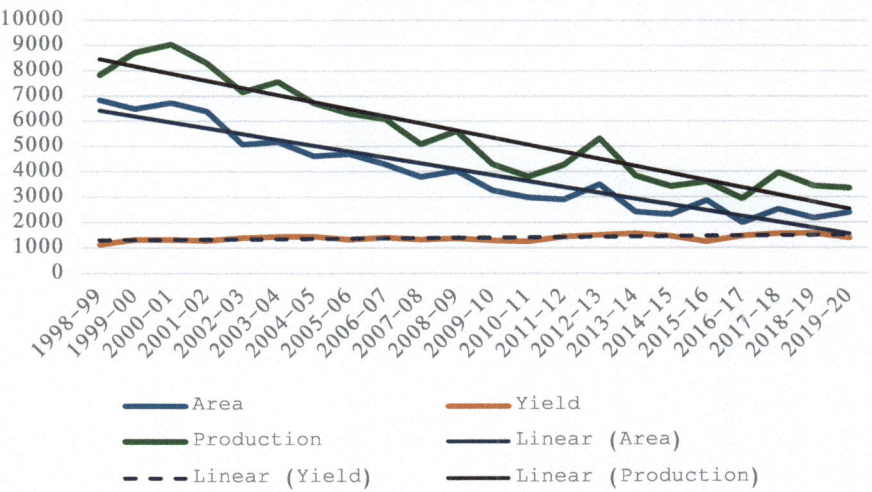

Fig. 5.1 Area, yield, and production of rice in Kozhikode between 1998 and 2020

Fig. 5.2 Economic considerations of social groups in Thamarakulam

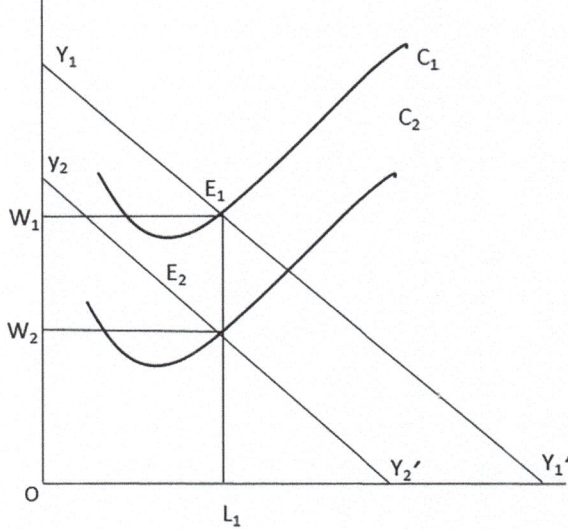

The terms of the lease would depend on the options available to both social groups. In regions where agricultural labour has very low levels of education, and hence not too many options outside agriculture, the terms of the lease could well be more in favour of the landowner seeking to cease cultivation.

Chapter 6
Sliding Plantations in Anappara

The Anappara village, located on the foothills of the Western Ghats, has experienced both long-term and extreme events and their multiple impacts in the last two or more decades. The uncertain rainfall events—unseasonal and extreme—have deteriorating effects on the plantation crops that contribute to a significant share of the village's local economy. Plantation crops such as Coconut, rubber, tea, areca nut, coffee, oil palm, areca nut, cashew, and cocoa are susceptible to the variability in rainfall and humidity.[1] Unseasonal rainfall, delayed monsoons, dry spells during the rainy season, and heavy rainfall of short duration alter the production cycle of coconut and areca nuts.[2] The rainfall variability experienced by Anappara has decreased crop productivity for rubber, coconut, and areca nuts. The reduction in rubber productivity is attributed to the decrease in tapping days due to heavy and extreme rainfall events.[3]

In Kozhikode, the heavy rainfall causes severe button shedding in coconut trees, which results in less production.[4] Likewise, areca nut production is susceptible to variability in daily temperature, relative humidity, rainfall, and sunshine hours. Rainfall during the fruit development stage and evening relative humidity significantly negatively impact areca nut yield. The big and small cultivators who depend on various plantation crops are the primary victims of these impacts of the changing climate.

---

[1] Hebbar et al. (2016).
[2] Khandekar et al. (2020).
[3] Ruangsri et al. (2015).
[4] Abhinav et al. (2018).

6.1 The Impact of Climate Change on Plantation Crops

As far as people's experiences are concerned—particularly those who have been dependent on plantation crops, the changing climate has multiple impacts on rubber, cocoa, and coconut-major crops produced in Anappara. The variability in rainfall and temperature often leads to pests and plant diseases. The incidence of pests and plant diseases is high in Coconut, Areca nut, rubber, pepper, and cocoa. The heavy rainfall spell leads to waterlogging, a lower rate of aeration and evapotranspiration, fewer sunshine hours, and decreased nutrition intake, which increases the incidence of pests and disease in coconuts in the Kozhikode region.[5]

The climatic conditions directly impacted the crops' phenology and, thus, the crop growth, yield, and production. Apart from that, the changes in the climatic conditions also determine and sometimes facilitate the proliferation of pest and crop diseases. The incidence of crop disease has increased significantly in Anappara. The areca nut faced a huge threat from the fruit rot disease locally known as "Mahali." In Coconut, it was wilt disease, and in cocoa, it is a stem cranker.[6] In the early 1990s, the areca nut plantation in the village experienced the spread of Mahali disease. It eventually ruined all the areca nut cultivation in the village. coconut, pepper, and cocoa also had multiple diseases. The variation in temperature, rainfall, and relative humidity significantly impact the spread of crop diseases in coconut,[7] areca nut,[8] and cocoa. The fungus-like microorganism *Phytophthora palmivora* infected areca nut, coconut, and cocoa plantations in the district.[9]

6.1.1 Jose—The Long-Standing Planter of Anappara

The experiences of Jose, one of the big planters in Anappara, revealed the plights of dealing with the multiple impacts of changing climates on the plantations. Jose owes 50 acres of the plantation, where he cultivates multiple crops such as rubber, cocoa, coconut, and areca nuts. Jose is considered a second-generation planter in Anappara. His father had arrived in the village as a manager at the rubber plantation owned by Murikkan, one of the three big planters who began plantations in Anappara in the mid-twentieth century. The decline of the bigger plantations in the post-land reform era benefited Jose's father, who had 68 acres of land in the village. Jose's father settled in the village and became a big planter among the other small planters who emerged as a result of the departure of the big planters in the village. Over 68 acres of land provide Jose's family economic opportunities and status. His father has a high social status in the village. Several households in the village depended only on the

---

[5] Abhinav et al.
[6] Prabha and Chandramohanan (2011).
[7] Abhinav et al. (2018).
[8] Patil et al. (2022).
[9] Mohanan (2011), Patil et al. (2021), Syamsuddin et al. (2021).

labour in the plantation owned by Jose's father. While growing up, he saw people coming to his house to see his father and get plantation jobs. Jose was fascinated by his father's charisma as a big planter in the village.

After completing his post-graduate studies in botany in the 1980s at the University of Calicut, Jose entered the plantation with a great aspiration to become like his father. He was the only one in his university batch to enter cultivation. No one in his family resisted his decision to enter cultivation. One of the reasons, he recalled, was that cultivating was more profitable than any other job available in and around the village at that time. Rubber, coconut, area nut, nutmug, and coffee were the major crops in the village when he entered the plantation as a cultivator. Over the last four decades, Jose has witnessed many transformations in the plantation field. When he began the cultivation, one major issue that many planters in the village were experiencing was the spread of diseases in Areca nut.

In the late 1980s and the early 1990s, many cultivators who depended on the Areca nut lost their crops and ended up with bank loan debts and defaults. Jose, too, had six acres of Areca nut plantation. After losing some of his areca nut plants due to the spread of disease, he cut all of them and planted Cocoa plants. Cocoa plantation was the trend of that time and was even more profitable than the Areca nut. By the end of the 1990s, similar issues had happened to pepper, a crop that people had even cultivated on the homestead. Like areca nut, the pepper began to vanish from the village by the late 1990s. Rubber, Cocoa, and Coconut are the major crops that Jose has depended on in the last two decades.

Jose believe that plantation crops are no longer a viable economic option considering the current situation. He never encouraged his two sons to enter the plantation; they were well-educated enough to find employment in Dubai and the United Kingdom, respectively. Jose, at present, has the biggest crisis in keeping his plantations. The spread of diseases and crop failure in recent years has become very common in rubber, cocoa, and Coconut. The increase in the wage inputs and decline in the market price of these crops further multiplied the crisis. Jose explained the emotionally changing scenario of the plantation sector in the village.

> Earlier, we had no difficulties finding labourers on the plantation. People used to approach us seeking employment opportunities. Giving opportunities to all was the difficult thing. Now, I have been going to people's houses, seeking them to come to do work on the plantation. However, finding a labourer is also not going to solve the crisis. At the end of the day, what we get from the plantation is only sufficient to provide wages to the labourers. In the last few years, I have done the taping and other work in the rubber plantation around my house. Only then can we get something out of it'

6.2 The Plights of Seasonal In-Migrant Beekeepers

The changes in weather events and their subsequent impacts have affected not only the plantation cultivators in the village but also the seasonal migrant bee-keepers from the neighboring state. The cohort migrant mostly belongs to the Christian Nadar community from Marthandan of Kanyakumari district of Tamil Nadu are in migrants to Anappara since early 1990s. They travel 450 km from Kanyakumari and stay at Anappara for three to four months every year, from December–January to April–May. During their stay, they have informal arrangements with the local plantation owners and put their bee boxes across the plantations in the Anappara. In January–February, it is the time of leaf fall of from rubber tree and the sprouting of the new leaf and flowering. In the last four decades, the giant rubber plantation area has provided a stable economic opportunity for the seasonal migrants who only gathered honey in the new leaf and flowers of the rubber tree.

In the last few years, the seasonal migrants have experienced a severe decline in the honey they could collect. They, too, had to blame for the inappropriate weather events. Ashokan, a 50-year-old beekeeper in the group who has been visiting Anappara every year since his early 20s, observed that—the continuous rain in January and February was not expected. However, it has happened in the village in the last few years. This heavy rain wipes all the honey from the newly sprouted leaf, reducing the amount of honey and severely affecting the health of bees. In 2022, the season was disastrous for the beekeepers in Anappara. In March, in the middle of the regular season, due to a lack of honey, for the first time in the village's history, the migrants were forced to move to Nilambur in Malappuram district, in search of appropriate site to place their bee-boxes. In recent years, changing weather patterns in Anappara have become unfavorable for beekeepers. The accumulation of losses over the past few years has brought an end to the three-decade-long relationship between the Kanyakumari beekeepers and the Anappara planters

6.3 Extreme Events in Anappara

Apart from variability in climate profile, the region also faces threats from extreme events. Steep slopes of the Western Ghats in Kozhikode are one of the landslide susceptible zones of the state. Anappara gram panchayath in Kozhikode, where our study area is situated, has more than 40% of its total area under the moderately volatile zones regarding landslide risk zonation. Various locations under the panchayath have recently experienced multiple slumps, slope failures, and sometimes debris flow.[10] These events have resulted in casualties and incurred heavy economic, social, and infrastructural damages.[11] This condition is the reason the village has become more prone to landslides.

In the last two decades, the Anappara has been witnessing an increasing number of devastating landslides. In 2009, along with the extreme rainfall events and subsequent landslides in the adjacent areas, the government issued an alarm in Anappara village; many villagers, especially those in high-risk areas, were temporarily shifted to rehabilitation camps. In 2018 and 2019, the landslide devastated the people's lives in the village. In 2018, one person died, a bridge connecting the settlement areas

---

[10] Sreekumar (2019).

[11] Kerala State Disaster Management Authority (2021).

to the nearby market was damaged, fifteen houses were damaged completely, and the landslide partially damaged 98 houses. Before the incident, people were rehabilitated to safer places, which saved many lives. However, due to the landslide in 2018, the government permanently relocated twenty households to a relatively safer place in the nearby villages. The landslide in 2019 further accelerated the fear of the villagers' lives, infrastructure, and livelihood.[12]

According to our survey, though 75% of the households lived in concrete-roofed houses (RCC) the frequent landslides have multiple impacts on the everyday lives of the people in Anappara. There is an increasing fear among the people in the village who inhabit the elevated areas. The rehabilitated households have another crisis, as many live far away from their plantation.

6.4 Economic Fragility of Anappara

Small planters in Anappara grew crops such as areca nut, Coconut, Cocoa, pepper, and rubber. In recent decades, small planters have experienced multiple challenges due to long-term changes in critical climatic parameters. These challenges include pest attacks, the spread of new diseases, and an eventual decline in productivity. This crisis was reflected in a decline in market price, an increase in input cost, etc.

Rubber was the major corps in the village that survived when all other crops declined due to the deceases. However, in recent times, rubber planters have been in a severe crisis due to the unseasonal rainfall in January when the plant sprouts new leaves. The rain makes the new leaves fall, deteriorating the tree's health and lactose production. The decline in milk production added to the rubber planters' crisis, which was already experiencing a decline in market price. The crop failure due to the multiple issues generated by the long-term changes in climate prompted multiple responses from the cultivators, including selling the land for commercial purposes- especially for mining and experimenting with different types of plants that had more market price at a particular point in time.

6.5 Changes in the Labour Relation

Even though the planters, in general, irrespective of the scale, have been experiencing the accumulated stress of all these crises, the responses differed across the small and big planters. Many of the small cultivators have already begun using family labour in their plantations, which is the only way they can tap the benefits of the plantation during the crisis. Family labour was not a viable, adaptable strategy for the bigger planters. They inevitably were forced to make informal arrangements with labourers. Instead of wages, the labourers get an equal share of the outputs.

---

[12] Kerala State Disaster Management Authority.

Varky, a 69-year-old planter in Anappara, has 350 rubbers in his plantation attached to his home. He employed labourers who taped and processed the rubber lactose to the rubber sheet for a long time. Until 2021, he gave two rupees per rubber plant as a wage for taping, fifty paise per sheet. Together, it costs him rupees 800 per day as labour is charged alone. His revenue per day was nearly 1000 after transportation costs. In addition, he has experienced the result of the rainfall event badly in the past years. Earlier, in a season, he used to tape 100 days, but it has been reduced to around 60 tapings per year. Beyond the decline of the taping days, the rainfall event also resulted from low milk productivity. From 2021 onwards, Varky would do taping and other related jobs on his plantation. Varky found the wage that by saving through employing his labour, he could sustain the plantation. It was not the case for Varky alone. The movement of small planters entering their plantations as labourers was initiated in the village nearly a decade ago. Varky found himself as the last among the planters who did this.

What Varky did was impossible for Mathew, a 55-year-old big planter in the village. Mathew had 1500 rubbers, five acres of Coconut, an aeronaut, and so on. Getting to the plantation is not a possible solution for Mathew, though he seems a hard worker. He alone cannot complete a taping of 1500 every day. The rubber tapers on his plantation now made an informal leasing arrangement, sharing the output equally with Mathew. Mathew doesn't entirely like these because he believes that now the labourers can exploit the rubber, which will affect the plant adversely in the long run. However, there was no other option for him. Share of output is also a new arrangement that bigger planters are now moving towards in Anappara.

6.6 Reversal of Migration

The environmental fragility of the landscape was exposed to the effects of climate change, which triggered economic fragility and paved the way for the movement of small planters out of Anappara. Like any other hilly terrain, the Western Ghats tend to become fragile due to unscientific development activities; the land use land cover change and heavy rainfall episodes have made it vulnerable to landslides. The changing climate had become a significant force that generated despair among the cultivator, leading to their detachment from the land.

The slopes of Anappara, which historically attracted planters towards it due to its enabling conditions to plant rubber, had been forcing them out of it due to changing conditions. In the 1940s and 1970s, the economic crisis that people experienced in the plains was the factor that prompted the cultivators to climb the hills. The availability of a larger area of land at a cheaper price, the fertility of the land, and the hard work of the new cultivators indeed made Anappara a full-fledged village in a short period. Until the early 2000s, those with one or two pieces of land could lead a sustainable life in the village. Though the long-term climate change impacted the production of the crops the recurring extreme events triggered the movement out of the village fearing the loss of life and property. Those who were financially better off adapted to the crisis by moving out of the plantation agriculture, and they migrated from the hills to the plain. Those who left the hills at the present are the most economically disadvantaged sects in the village. The economically well-off families from the dominant groups of Mappila Muslims and Christians now lived in relatively safer places down the hills, and they leased out their land on the hills to the labourers, who received a share of the crops. Those who are left in the hills are largely the labourers who now manage their cultivation in small areas and the cultivation of others who already left the hills. This movement from hills to plains induced by the Climate crisis represents the reversal of the in-migration which formerly led to the creation of Anappara. Those who moved out of cultivation and found livelihood in town areas in the plain now found it difficult to stay in the hilly part of the village.

According to our Household survey (2022), in Anappara, the two most prominent communities are the Mappila Muslims and the Christians. Their Degree of Dominance (DID)[13] stood at 0.46 and 0.33, respectively. Even in terms of population, 51.10% of our sample population constituted by Mappila Muslims, 28.45% by Christians, and the remaining 17.21% by Hindus. These two communities also held more than 80% of the total plantation or agricultural landholdings in our sample; the Christians of Anappara owned 66.98% of the total landholdings, while the Mappila held 26.8%. Even the percentage of cultivators and labourers represented by these two communities. 75.76% of the cultivators from our sample were represented by Christians and 15.15% by Mappila Muslims. In terms of plantation labour as well, these two communities were found to constitute 43.75% of the plantation labour force in our sample was represented by Christians and 46.88% by Muslims. These results also suggest the transformations in land relations where the planters have to become labourers in their own plantations, while the labourers diversified their livelihoods by seeking opportunities in the nearby urban centres.

6.7 Urbanization and Opportunities

The declining scope of the plantation sector also forced the cultivators to look for other opportunities besides the rearrangements in the labour and the capital. The occupation profile analysis from the survey data revealed that among the working-age group from our sample, 7.07% of them identified themselves as planters/cultivators, 6.85% as plantation labourers while most of them were either employed as skilled workers (18.63%) including Driver, electrician, barber, salesperson etc., or other wage labourers (17.34%). A decent proportion of the working-age population was also employed in Private salaried Jobs (13.28%).

One notable opportunity that some people in the village have explored in the past two decades is the conversation of the plantation land for commercial purposes. The

---

[13] Pani (2022). The Degree of Identity Dominance (DID) involves average of the indicators of the three instruments such as the Economic power of identity group, Political dominance of identity group and dominance of an identity group in the realm of knowledge.

establishment of the Aided Collage in 2003 by the Church initiated an urbanization process in the village. The college offers undergraduate and post-graduation degrees in social work, psychology, English, and journalism. Nearly 1000 students from across the district in Kerala are perusing these courses. It generated a new economic opportunity for homestays, guest houses, restaurants, and shops in the village. This resulted in the conversation of the plantation for commercial purposes. However, only some were able to tap these new opportunities similarly. The big cultivators and the gulf migrants were the two sects in the village who could tap these opportunities.

The availability of granite stone in the village also provided new opportunities for the landowning planters in Anappara. The distress in plantation agriculture and the high demand for granite stones forced villagers to tap this new opportunity. Here, it is different from the people who are experiencing the crisis and who have directly moved to convert the land; instead, in most cases, the granite quarries are managed by people from outside the village. The failure of the crops and high returns from the mining has forced the cultivators to lease out the land for mining.

Chapter 7
External Interventions in Autonomous Adaptation

Autonomous adaptation to climate change is a less explored facet in the climate change adaptation discourse. An inquiry into the forms of adaptation that take place autonomously and spontaneously was initiated with an understanding of the exposure to multiple climate events induced by climate change as the characteristics of the exposure also determine the adaptation. Further exploration into the everyday adaptations to the changing climate at the individual level in one of the most exposed districts in the country led us to the observation that how cumulated everyday responses towards the effects of climate change involve the nexus between the environmental and socio-economic as well as the political process. The interactions between these processes while dealing with the risks imposed by climate change, especially on nature-based livelihoods such as fishing and agriculture. This also exposed the existing inequalities in the existing social structures and to different forms of social transformations. In the coastal village of Melthura, the catch of primary fish varieties took a toll due to the warming oceans and rising sea levels had forced them to change from where and how they carried out their fishing. These modifications made the fishing community change the way the boat owners and the labours shared their risk. In addition, it polarized the community based on the source of investments and brought about animosities, especially on the coast rather than in the seas.

In the paddy cultivating Thamarakulam, the rainfall variability and water management decisions inflicted indirectly by climate change amended the traditional land relations. The recurring risks forced the traditionally land-owing communities which could not gain the formerly existing level of profits to lease their lands to the labours who were ready to manage the lands without looking for large margins of profits. In the third site of study, the village of Anappara, when the changing rainfall and temperature made the plantations un-profitable due to pests and plant diseases. The repeated episodes of landslides triggered the migration out of the hills into the plains and thus accentuated urbanization. As already discussed the planned adaptation discourse overlooks the adaptation that is taking place at the individual and local level. So in the following sections based on our observations from our study area, we suggest possible site-specific interventions that could cater to more sustainable processes of changes in the era of climate change.

In terms of the interventions, we tried to discuss the issue that requires interventions, already existing local responses along with the possible future interventions and the authorities of implementation of the suggested interventions.

7.1 Melthura—The Abandoned Fishing Beach

In Melthura the notable issues involved the livelihood loss of the older fishermen; the young men from fishing households wanting to move away from fishing and the women who lost their jobs at the harbour. It also faced the issue of incoming migrant fishermen. Older fishermen in the village have lost their livelihood due to climate change as they cannot move to the new harbour in the nearby town to carry out fishing daily, especially during the fishing seasons. As a response to this, some of them stayed unemployed while a few others settled for off-season shore-based employment such as fishnet repair. Some of them even looked for other wage labour outside fishing. The Department of Fisheries already has the option of providing pensions to older fishermen, but the fishermen often feel inadequate.

When some of the fishermen had to move out of the fishing, it caused a lot of economic stress on the households. The women in the village who did odd jobs in the harbour had lost their livelihood with the local harbour becoming defunct. The women try to find other wage labour or work as saleswomen in the local shops or bakeries. Some of the women who were not allowed to work found this as a newfound opportunity for financial independence while aiding their families in distress. But not all of them find it easy to find employment within the close vicinity of the village. The women in the village could be employed by organizing themselves into local units. This can be realized by implementing flagship schemes like the Theeramythri programme[1] under the Society for Assistance of Fisherwomen functions under the Kerala Fisheries Department focuses on facilitating the self-employment of fisherwomen by encouraging them to organize themselves into small activities-based groups. The Programme extends financial, technological, and other managerial support to build up micro-enterprises. Similarly, Neighbourhood groups[2] of women can be created under the Kudumbasree scheme.[3] Here the state would be a major form of authority that could facilitate this intervention.

The arrival of the migrant fishermen to carry out line fishing to catch Ayakoora off the coast of Melthura has been causing hostility between the local and the migrant fishermen. These fishermen stay in the abandoned old harbour in their makeshift tents and after the season ends they move out of the village. However, this has increased the animosity between the local fishermen and the local middlemen who aid the migrant fishermen. To decrease this tension an effective intervention could involve the local

---

[1] Society for Assistance To Fisherwomen (SAF).

[2] Kudumbashree|What Is Kudumbashree.

[3] Shyam and Geetha (2013).

fishermen's welfare society or cooperative creating the required infrastructure that would facilitate the stay and operations of the migrant fishermen and by charging them for the services provided. This intervention would provide a form of revenue to the local cooperative which could be invested in the schemes that ensure the welfare of the local fishermen households. In addition, this would also ensure that the profit will not be vested in the hands of the specific individuals.

The precarity in fishing in recent times aggravated due to climate change made the younger generation look for other employment options, most of them look for options to migrate to other towns or gulf countries. Moving out of fishing and finding livelihood opportunities in non-fishing sectors within the vicinity of the village or far away places are not very easy for the younger generation of the fishermen's families. There are multiple forms of inequalities in terms of the distribution of economic and social capital among the youths of two different fishing communities to grab livelihood opportunities outside of fishing. For example, there is a solid inequality in terms of the distribution of migration capital and the connectivity to the migration network that is available for both the fishing communities in the village. This inequality, as the study revealed, in the regime of climate change distress, has manifested to the exacerbation of the polarization of these two communities.

As compared to the old generation, among the youths in the village, there is less experience of secular friendship ties. The Mukkuva youths have great feelings of being left behind in the opportunities while seeing the Puslan youth have a more economically and socially prosperous life. This situation indeed demands an intervention from all levels. At present, it seems that the communal organization and the political parties are the beneficent of the growing communal tensions in the village. Therefore, there is a need for interventions from the home, communities, and the state to reduce the inequalities among the youth of different social backgrounds who aspire the livelihood opportunities outside the fishing within or outside the village.

The arrival of the larger boat and the consolidation of the identity capital has led to more polarized groups, there has been less cooperation among the identity groups in the village. As an intervention to this segregation, NGOs involved in the welfare of fishermen could lead the formation of more secular fishing units by creating awareness of the advantages of community-level cooperation in sharing the climate risk.

7.2 Thamarakulam—The Flooded Rice Bowl

In Thamarakulam the vital points of interventions were found to have to be required in terms of flooding of the paddy fields, seed choices of the farmers, facilitation of the fish culture, and migration. As we understood, apart from floods from unseasonal rainfall drowning the crops, the paddy fields are also forcefully flooded to maintain the levels of the groundwater that feeds the drinking water supply network. Internventions from the state would be required to ensure the sustainability of the paddy cultivation in the region which forms the part of the "rice bowl of north Kerala". The state can consider creating infrastructure to supply drinking water from the Kuttiyadi dam to the village and the surrounding areas. Such an infrastructure has already found a place in the dam, a JICA-assisted Kerala Water Supply Project that caters to the drinking water supply to Kozhikode city and adjoining villages.[4] The state could either expand or create a new infrastructure. This would also decrease the pressure on the groundwater and give time for the aquifers to replenish naturally instead of the artificial recharge.

Unable to abide by the traditional climate-Agri calendars due to the shift in the seasonality caused by climate change, some of the farmers retracted back to the traditional seed varieties which either had flood or drought resistance. However, the availability of these seeds was observed to be a major issue. The farmers from Thamarakulam pinpointed that often in the state-run seed banks, the seeds preferred by them are not made available and are being sold whatever is available in the existing stock of seeds. One possible intervention to ensure the agency of the farmers in the choice of seeds is that the local farming cooperative societies could maintain the stock of traditional seed varieties and disburse it to the farmers at a discounted price. These societies could even provide incentives to the members who contribute to the seed banks.

When the paddy fields remained flooded for an extended period and as a result, the cultivation could not take place some of the farmers saw this as an opportunity to grow fish in their water-filled lands. The integrated rice-fish farming which was traditionally practiced in the Northern Part of Kerala requires special skills and awareness of the techniques and processes.[5] However, the farmers in Thamarakulam were not successful as most of them lacked awareness of techniques of fish culture and these ventures ended up unproductive without any profit. The state can intervene and provide the necessary training and financial mechanisms to create infrastructure to grow fish in the agricultural lands. The Agency for the Development of Aquaculture, functioning under the Kerala Fisheries Department has been involved in the implementation of the schemes related to the integrated development of inland fisheries and aquaculture.[6] This particular agency could assess the scope for the improvement of aquaculture in the village and provide the necessary assistance to the farmers interested in involving in the fish culture.

It is also a reality that the Thamarakulam paddy field is no longer a viable livelihood opportunity for the village people. The growing movement out of agriculture, migration, and livelihood diversification trends in the villages, captured in the study are the inevitable result of the distress in agriculture and also the growing opportunities in the outside. However, there is also a profound inequality between the members of different communities in the way the capital and networks available to opt for these coping strategies. Multiple agencies and institutions are already working at government and aided levels providing skills to the specific age, gender, and identity groups to attain livelihood opportunities outside the agriculture. The role of this initiative needs to be evaluated in the context of the increasing crisis in the local economy and the need to plan adequate strategies to help a diverse set of people across age, gender, and social identities tap the opportunities outside agriculture.

---

[4] PASK—KWA Project Monitoring System.

[5] Sathoria and Roy.

[6] Inland Fish Production|Fisheries Department—Kerala.

7.3 Anappara—The Sliding Plantations

The village of Anappara situated on the Western Ghats, was surrounded by issues related to plant diseases and pest attacks apart from extreme events like landslides. As we had observed in the chapter on the impacts of climate change in Anappara, the changing patterns of rainfall and temperature had impacted the productivity of the major crops such as Rubber, Arecanut, and Coconut. As the large plantation owners could not maintain profits they sold their land to the small planters who were ready to take risks. But the recurring landslides the recent times have triggered the fear regarding the loss of life and property among the planters in the village. Most of them want to relocate to safer locations but they are tied to their lands. One efficient intervention to deal with the issue could be for the planters could collectively hire a private cooperative which involves people who would manage their lands at a price. This would allow the planters to relocate to the nearby towns while enjoying a fair share of the earnings from the plantations. In addition, they can also find employment in other sectors which would improve the overall earnings of their households.

Anappara is a case to understand what is happening in the plantation sector in the country. Some of the crises are specific to the local context and some have larger patterns as well. The spread of new diseases, a decline in productivity, and subsequent migration of the people who depended on the plantation crops to the non-farm sector have been not a story of Anappara alone. In such cases, there is an urgent intervention required from the states to protect the sustainability of the plantation economy of the country.

One of the urgent and broad interventions at the state level is to strengthen the research and development in the plantation sector- specifically to find adequate solutions to the spread of new diseases. Since the plantation crops have significant scope in the larger industrial productions, and economy of the country, it seems important to address the fundamental issues of the climate change that has been affecting plantation crops and the people who depend on it. Setting up research institutes at the local level and promoting applied scientific research in this area is important to bring sustainability to the plantation sector.

Specific interventions are also needed to address the issues that are specific to the local context of the Anappara. One such issue is the plight of the cultivators, who are relocated to places far from the village, as a strategy to protect them from the threat of extreme events- specifically the landslides. The present study revealed that the household that relocated five km away from the village has cultivable land in Anappara and many of them are small planters. These relocated small planters are now facing difficulties in a way that they have to put in additional efforts and spend money to reach the village to engage in cultivation. These issues need more attention at the local village level where civil society groups, NGOs, and the state can make new interventions to help the relocated people to reduce their input cost and maintain the cultivation.

7.4 Conclusion

In recent years, the importance of autonomous adaptation to climate change—and the need to incorporate its insights and strategies into planned adaptation—has become a growing concern in the field of adaptation governance. However, the convergence of planned and autonomous adaptation remains a complex and unresolved challenge. This issue has gained significant attention in both academic research and policy-making circles. Unlike planned adaptation, autonomous adaptation is highly diverse in form and is often shaped by a range of social, cultural, political, and ecological dynamics. As such, there is a pressing need to explore how autonomous adaptation unfolds across different geographies, cultures, and political contexts, with particular attention to the nature and degree of exposure to climate change variables. It is in this context that this book presents the story of how the intersection of autonomous adaptation and inequality is driving social, economic, and political transformations across the coast, plains, and hills of Kozhikode—one of India's districts most exposed to multiple climatic events over the past three decades. As we witnessed how the issues are differential at the individual and local level and the suggested interventions require the state as one of the major sources of authority along with the Non-Governmental Organizations it is evident that the planned responses towards coping the climate crisis must consider understanding the processes involved in the autonomous responses to the direct and indirect impacts of climate change. It is observed that the welfare schemes and community development projects for the fishing communities are often planned at the state level and implemented among the fishermen through the district bodies. This approach indeed lacks the complex dynamics of the local issues. Specifically in the time of climate change the issues experienced by each village and their resilience capacity may vary. Similarly, the waterlogging in the paddy fields of Thamarakulam, which deteriorated the village's paddy cultivation, was a consequence of maladaptive planned policies aimed at solving the drinking water crisis in the village. The state-led rehabilitation of landslide victims in Anappara separated cultivators from their land, leading to multiple crises later. Therefore, it is important to understand what brings a sustainable solution to the problem in the local context. A collaboration between the local government institutions, local people, local civil society groups, and experts in the field in planning and implementation can provide better results.

The need to converge autonomous adaptation with planned adaptation calls for a shift in the discourse on climate change adaptation towards the study of democratic innovations. Otherwise, discussions on climate change—particularly those concerning the governance of adaptation—must be integrated with existing debates on decentralisation planning and buddgeting, participatory democracy, and deep democracy which are crucial for addressing climate crises at the local level and reducing the risks of maladaptation

References

A Timeline of the Local Communities and Indigenous Peoples Platform|Local Communities and Indigenous Peoples Platform. Accessed 31 Jan 2024. https://lcipp.unfccc.int/lcipp-background/timeline-local-communities-and-indigenous-peoples-platform

Abhinav MC, Paul Lazarus T, Priyanga V, Kshama AV (2018) Impact of rainfall on the coconut productivity in Kozhikode and Malappuram districts of Kerala. Curr Agricult Res J 6(2): 183–87. https://doi.org/10.12944/CARJ.6.2.07

Abhinav MC (2018) Impact of rainfall on the coconut productivity in Kozhikode and Malappuram districts of Kerala. Curr Agricult Res J 6(2): 183–87. https://doi.org/10.12944/CARJ.6.2.07

Adve N (2014) Moving home: global warming and the shifts in species' range in India. Econ Polit Weekly 49(39): 34–38. https://www.jstor.org/stable/24480732

Ambli S (2021) Major land reform legislations in Kerala. Int J Creat Res Thoughts 9(12):230–236

Aune KT, Davis MF, Smith GS (2021) Extreme precipitation events and infectious disease risk: a scoping review and framework for infectious respiratory viruses. Int J Environ Res Public Health 19(1):165. https://doi.org/10.3390/ijerph19010165

Bannayan M, Sadeghi Lotfabadi S, Sanjani S, Mohamadian A, Aghaalikhani M (2011) Effects of precipitation and temperature on crop production variability in Northeast Iran. Int J Biometeorol 55(3):387–401. https://doi.org/10.1007/s00484-010-0348-7

Banuri T, Barker T, Bashmakov I, Blok K, Bouille D, Christ R, Davidson O, Edmonds J, Gregory K, Grubb M, Halsnaes K, Heller T, Hourcade J-C, Jepma C, Kauppi P, Markandya A, Metz B, Moomaw W, Moreira JR, Morita T, Nakicenovic N, Price L, Richels R, Robinson J, Rogner HH, Sathaye J, Sedjo R, Shukla P, Srivastava L, Swart R, Toth F, Weyant J (2001) Climate change 2001: mitigation: contribution of working group III to third assessment report of intergovernmental panel on climate change. IPCC, Geneva. https://www.ipcc.ch/site/assets/uploads/2018/03/WGIII_TAR_full_report.pdf

Bedane HR, Beketie KT, Fantahun EE, Feyisa GL, Anose FA (2022) The impact of rainfall variability and crop production on Vertisols in the central highlands of Ethiopia. Environ Syst Res 11(1):26. https://doi.org/10.1186/s40068-022-00275-3

Biagini B, Kuhl L, Gallagher KS, Ortiz C (2014) Technology transfer for adaptation. Nat Clim Change 4(9):828–834. https://doi.org/10.1038/nclimate2305

Buonocore M. What is climate? What is climate change? Climateurope (blog). Accessed 6 Feb 2024. https://www.climateurope.eu/what-is-climate-and-climate-change/

Burton I (1997) Vulnerability and adaptive response in the context of climate and climate change. Clim Change 36(1):185–196. https://doi.org/10.1023/A:1005334926618

Cai WH, Yang YZ, Yang J, He HS (2018) Topographic variation in the climatic change response of a larch forest in Northeastern China. Landscape Ecol 33(11):2013–2029. https://doi.org/10.1007/s10980-018-0711-3

Chattopadhyay N, Sahai A, Guhathakurta P, Dutta S, Srivastava A, Attri SD, Ramamurthy B, Malathi K, Chandras S (2019) Impact of observed climate change on the classification of agroclimatic zones in India. Curr Sci 117:480

Cutter SL (2018) Compound, cascading, or complex disasters: what's in a name? Environ Sci Policy Sust Develop 60(6):16–25. https://doi.org/10.1080/00139157.2018.1517518

DES|Area, Production and Yield—Reports. Accessed 15 Mar 2024. https://data.desagri.gov.in/website/crops-apy-report-web

Directorate of Census Operations, Kerala. District census handbook—village and town directory, Kozhikode. District Handbook. Kerala: Census of India 2011, 2014. https://censusindia.gov.in/nada/index.php/catalog/656

Field CB, Barros VR, Panel I on Climate Change (eds) (2014) Climate change 2014: impacts, adaptation, and vulnerability: Working Group II contribution to the fifth assessment report of the intergovernmental panel on climate change. Cambridge University Press, New York

Floods|NDMA, GoI. Accessed 22 Feb 2024. https://ndma.gov.in/Natural-Hazards/Floods

Fund, Green Climate Portfolio Dashboard. Text. Green Climate Fund. Green Climate Fund. Accessed 30 Oct 2023. https://www.greenclimate.fund/projects/dashboard

George R, Syda R (2012) Marine fishery development and climate change. Accessed 1 Jan 2012

Global Climate Observing System (GCOS) (2016) Indicators of climate change. World Meteorological Organization. http://ane4bf-datap1.s3-eu-west-1.amazonaws.com/wmocms/s3fs-public/ckeditor/files/15.1_GCOS_climate_indicator_summary.pdf?Xeq3ylSFE_AsgkS55oiXi8glMywGjgx3

Global Facility for Disaster Reduction and Recovery (GFDRR) Think hazard—India. Think Hazard. Accessed 13 Aug 2021. https://thinkhazard.org/en/report/115-india

Gopinath G, Ambili GK, Gregory SJ, Anusha CK (2015) Drought risk mapping of south-western state in the Indian Peninsula—a web based application. J Environ Manag 161:453–459. https://doi.org/10.1016/j.jenvman.2014.12.040

Gopinath G, Surendran U, Abhilash S, NagaKumar KCV, Anusha CK (2020) Assessment of drought with a real-time web-based application for drought management in humid tropical Kerala. India. Environ Monit Assess 192(11):728. https://doi.org/10.1007/s10661-020-08665-9

Greeshma MS, Jairaj PG (2014) Coastal vulnerability assessment along Kerala using remote sensing and GIS. Int J Sci Eng Res 5(7):7

Groundwater Department (2021) Groundwater level monitoring report—June 2021. Government of Kerala. https://groundwater.kerala.gov.in/wp-content/uploads/2018/04/Water-Level-Report-JUNE-2021.pdf

Guhathakurta P, Menon P, Inkane PM, Krishnan U, Sable ST (2017) Trends and variability of meteorological drought over the districts of India using standardized precipitation index. J Earth Syst Sci 126(8):120. https://doi.org/10.1007/s12040-017-0896-x

Gulati D, Satpute S, Kaur S, Aggarwal R (2022) Estimation of potential recharge through direct seeded and transplanted rice fields in semi-arid regions of Punjab using HYDRUS-1D. Paddy Water Environ 20(1):79–92. https://doi.org/10.1007/s10333-021-00876-1

Hamza F, Valsala V, Mallissery A, George G (2021) Climate impacts on the landings of Indian Oil sardine over the South-Eastern Arabian sea. Fish Fish 22(1):175–193. https://doi.org/10.1111/faf.12513

Hebbar K, Berwal M, Chaturvedi VK (2016) Plantation crops: climatic risks and adaptation strategies. Indian J Plant Physiol 21:428–436. https://doi.org/10.1007/s40502-016-0265-9

Holmes EE, Br S, Nimit K, Maity S, Checkley SM, Wells ML, Trainer VL (2021) Improving landings forecasts using environmental covariates: a case study on the Indian Oil sardine (*Sardinella longiceps*). Fish Oceanogr 30(6):623–642. https://doi.org/10.1111/fog.12541

Hussain SV, Ulahannan ZP, Joseph D, Somasekharan A, Girindran R, Benny S, Ninan RG, Valappil ST (2021) Impact of climate change on the fishery of Indian Mackerel (*Rastrelliger kanagurta*) along the Kerala coast off the Southeastern Arabian Sea. Reg Stud Mar Sci 44:101773. https://doi.org/10.1016/j.rsma.2021.101773

References

Indian Meteorological Department (IMD) (2016a) Annual climate summary—2015. Indian Meteorological Department, New Delhi

Indian Meteorological Department (IMD) (2016b) Extremes of temperature and rainfall for Indian stations (up to 2012). Ministry of Earth Sciences, New Delhi. https://imdpune.gov.in/library/public/EXTREMES%20OF%20TEMPERATURE%20and%20RAINFALL%20upto%202012.pdf

Indian Meteorological Department (IMD) (2021) Cyclone warning in India standard operation procedure. Ministry of Earth Sciences, New Delhi. https://mausam.imd.gov.in/imd_latest/contents/pdf/cyclone_sop.pdf

Indian Meteorological Department (IMD). Observed rainfall variability and changes over different state. Climate Research and Services, Pune. Accessed 12 Dec 2021. https://www.imdpune.gov.in/reports.php

Indian Meteorological Department (IMD) (2020) Heat wave warning services. New Delhi. https://mausam.imd.gov.in/imd_latest/contents/pdf/pubbrochures/Heat%20Wave%20Warning%20Services.pdf

Indian Meteorological Department. Climate research and services, Pune. Statewise Temperature Trends. Accessed 1 Mar 2024. https://imdpune.gov.in/climinform.php

Inland Fish Production|Fisheries Department—Kerala. Accessed 21 Mar 2024. https://fisheries.kerala.gov.in/index.php/en/inland-fish-production

IPCC (2021) Climate change 2021: the physical science basis. In: Contribution of working group I to the sixth assessment report of the intergovernmental panel on climate change. Cambridge University Press. In Press: Intergovernmantal Panel on Climate Change

IPCC (2022) IPCC glossary search. Accessed 14 Mar 2022. https://apps.ipcc.ch/glossary/

Irwandi H, Rosid MS, Mart T (2023) Effects of climate change on temperature and precipitation in the Lake Toba region, Indonesia, based on ERA5-land data with quantile mapping bias correction. Sci Rep 13(1):2542. https://doi.org/10.1038/s41598-023-29592-y

Kankara RS, Ramana Murthy MV, Rajeevan M (2018) National assessment of shoreline changes along Indian coast. Assessment report. National Centre for Coastal Research, Chennai. https://www.indiaspend.com/wp-content/uploads/2018/11/National-Assessment-of-Shoreline-Changes-NCCR-report.pdf

Kerala State Disaster Management Authority (2021) Event report on 'extreme rainfall over Kerala' October 2021. Government of Kerala. https://sdma.kerala.gov.in/wp-content/uploads/2022/05/Event-report_October-2021.pdf

Khandekar RG, Desai V, Dhopavkar R, Borkar P, Haldankar P, Arulraj S, Parulekar Y et al (2020) Coconut and arecanut production protocol under aberrant climate of coastal region. Adv Agricult Res Technol J 4:97–104

Kripa V, Mohamed KS, Said Koya LP, Jeyabaskaran R, Prema D, Padua S, Kuriakose S et al (2018) Overfishing and climate drives changes in biology and recruitment of the Indian Oil sardine *Sardinella longiceps* in Southeastern Arabian sea. Front Mar Sci 5:443. https://doi.org/10.3389/fmars.2018.00443

Krishnan R, Dhara C (2020) Executive summary. In: Assessment of climate change over the Indian region. Springer Open, Singapore

Krishnan R, Sanjay J, Gnanaseelan C, Mujumdar M, Kulkarni A, Chakraborty S (2020) Assessment of climate change over the Indian region a report of the ministry of earth sciences (MoES), Government of India. https://doi.org/10.1007/978-981-15-4327-2

Krol M, Jaeger A, Bronstert A, Güntner A (2006) Integrated modelling of climate, water, soil, agricultural and socio-economic processes: a general introduction of the methodology and some exemplary results from the semi-arid north-east of Brazil. J Hydrol 328(3–4):417–431. https://doi.org/10.1016/j.jhydrol.2005.12.021

Kudumbashree|What Is Kudumbashree. Accessed 20 Mar 2024. https://www.kudumbashree.org/pages/171

Kurien J (1985) Technical assistance projects and socio-economic change: Norwegian intervention in Kerala's fisheries development. Econ Polit Weekly 20(25/26):A70-88

Lawrence A, Hoffmann S, Beierkuhnlein C (2021) Topographic diversity as an indicator for resilience of terrestrial protected areas against climate change. Glob Ecol Conserv 25:e01445. https://doi.org/10.1016/j.gecco.2020.e01445

Luers AL, Lobell DB, Sklar LS, Lee Addams C, Matson PA (2003) A method for quantifying vulnerability, applied to the agricultural system of the Yaqui Valley, Mexico. Glob Environ Change 13(4):255–267. https://doi.org/10.1016/S0959-3780(03)00054-2

Marcinkowski P, Piniewski M (2018) Effect of climate change on sowing and harvest dates of spring barley and maize in Poland. Int Agrophys 32(2):265–271. https://doi.org/10.1515/intag-2017-0015

Met Office. Causes of climate change. Accessed 23 Feb 2024. https://www.metoffice.gov.uk/weather/climate-change/causes-of-climate-change

Mohanan C (2011) Occurrence and distribution of cocoa (*Theobroma cocoa* L.) diseases in India. In: Peter PK, Chandramohanan R (eds). https://www.academia.edu/8445245/OCCURRENCE_AND_DISTRIBUTION_OF_COCOA_Theobroma_cocoa_L_DISEASES_IN_INDIA_By_P RABHA_K_PETER_and_R_CHANDRAMOHANAN

Mohanty A (2020) Preparing India for extreme climate events. Council on Energy, Environment and Water, New Delhi. https://www.ceew.in/sites/default/files/CEEW-Preparing-India-for-extreme-climate-events_10Dec20.pdf

Mujumdar M, Bhaskar P, Ramarao MVS, Uppara U, Goswami M, Borgaonkar H, Chakraborty S (2020) Droughts and floods. In: Assessment of climate change over the Indian region, pp 117–141. Springer Open, Singapore

Naga Kumar KCV, Deepak PM, Basheer Ahammed KK, Rao KN, Gopinath G, Dinesan VP (2022) Coastal vulnerability assessment using geospatial technologies and a multi-criteria decision making approach—a case study of Kozhikode district coast, Kerala State, India. J Coast Conserv 26(3):16. https://doi.org/10.1007/s11852-022-00862-7

National Adaptation Programmes of Action|UNFCCC. Accessed 30 Jan 2024. https://unfccc.int/topics/resilience/workstreams/national-adaptation-programmes-of-action/introduction?gclid=EAIaIQobChMI1oTNvruEhAMV36pmAh2JFQ6ZEAAYASAAEgJDuPD_BwE

Pai DS, Nair S, Ramanathan AN (2021) Long term climatology and trends of heat waves over India during the recent 50 years (1961–2010). Mausam 64(4): 585–604. https://doi.org/10.54302/mausam.v64i4.742.

Pani N (ed) (2022) Dynamics of difference: inequality and transformation in rural India. Routledge, London, New York

Pani N, National Institute of Advanced Studies (Bangalore, India) (eds) (2022) Dynamics of difference: inequality and transformation in rural India. First South Asia edition. Routledge, Taylor & Francis Group, London; New York

PASK—KWA project monitoring system. Accessed 21 Mar 2024. https://pask.kwa.kerala.gov.in/project/profile/reference/04e299e28c5847efc6b384bd74d81e25ioaEfg

Patil B, Hegde V, Maheswarappa HP, Narayanaswamy H (2021) Phytophthora diseases of arecanut in India: prior findings, present status and future prospects. Indian Phytopathol 74:382. https://doi.org/10.1007/s42360-021-00382-8

Patil B, Hegde V, Sridhara S, Narayanaswamy H, Naik MK, Patil KKR, Rajashekara H, Mishra AK (2022) Exploring the impact of climatic variables on arecanut fruit rot epidemic by understanding the disease dynamics in relation to space and time. J Fungi 8(7):745. https://doi.org/10.3390/jof8070745

Peter PK, Chandramohanan R (2011) Occurrence and distribution of cocoa (*Theobroma cocoa* L.) diseases in India. J Res Angrau 39(4):44–50

Prathipati VK, Naidu CV, Konatham P (2019) Inconsistency in the frequency of rainfall events in the Indian summer monsoon season. Int J Climatol 39(13):4907–4923. https://doi.org/10.1002/joc.6113

Pugh D, Woodworth P (2012) Sea-level science: understanding tides, surges, tsunamis and mean sea-level changes. https://doi.org/10.1017/CBO9781139235778

Punya P, Kripa V, Padua S, Sunil Mohamed K, Nameer PO (2021) Impact of environmental changes on the fishery of motorized and non-motorized sub-sectors of the upwelling zone of Kerala, Southeastern Arabian Sea. Estuar Coast Shelf Sci 250:107144. https://doi.org/10.1016/j.ecss.2020.107144

Rawat S, Ganapathy A, Agarwal A (2022) Drought characterization over Indian sub-continent using grace-based indices. Sci Rep 12(1):15432. https://doi.org/10.1038/s41598-022-18511-2

Ray D, Behera MD, Jacob J (2014) Indian brahmaputra valley offers significant potential for cultivation of rubber trees under changed climate. Curr Sci 107(3):461–469

Research (UCAR), University Corporation for Atmospheric. Climate change: regional impacts \textbar center for science education. Accessed 16 Feb 2022. https://scied.ucar.edu/learning-zone/climate-change-impacts/regional

Roxy MK, Ghosh S, Pathak A, Athulya R, Mujumdar M, Murtugudde R, Terray P, Rajeevan M (2017) A threefold rise in widespread extreme rain events over central India. Nat Commun 8(1):708. https://doi.org/10.1038/s41467-017-00744-9

Ruangsri K, Makkaew K, Sdoodee S (2015) The impact of rainfall fluctuation on days and rubber productivity in Songkhla province. Int J Agricult Technol. https://www.semanticscholar.org/paper/The-impact-of-rainfall-fluctuation-on-days-and-in-Ruangsri-Makkaew/03238927ca6be3aac766e7513c89c1bc83ad7045

Sabu T, Vinod KV, Jobi MC (2008) Life history, aggregation and dormancy of the rubber plantation litter beetle, *Luprops tristis*, from the rubber plantations of moist south western Ghats. J Insect Sci 8:1. https://doi.org/10.1673/031.008.0101

Salim SS, Ramees Rahman M, Pushkaran KN, Musaliyarakam N, Soma S (2016) Fish marketing: a market structure analysis of Kozhikode and Alappuzha districts. Mar Fish Inform Serv 24:18–23

Sanjay J, Revadekar JV, Ramarao MVS, Borgaonkar H, Sengupta S, Kothawale DR, Patel J, et al (2020) Temperature changes in India. In: Krishnan R, Sanjay J, Gnanaseelan C, Mujumdar M, Kulkarni A, Chakraborty S (eds) Assessment of climate change over the Indian region. Springer Singapore, Singapore, pp 21–45. https://doi.org/10.1007/978-981-15-4327-2_2

Sathoria P, Roy B (2022) Sustainable food production through integrated rice-fish farming in India: a brief review. Renew Agricult Food Syst 37(5):527–535. https://doi.org/10.1017/S1742170522000126

Saud S, Wang D, Fahad S, Alharby HF, Bamagoos AA, Alabdallah NM, AbdElgawad H, Adnan M, Sayyed RZ (2022) Comprehensive impacts of climate change on rice production and adaptive strategies in China. Front Microbiol 13:59. https://doi.org/10.3389/fmicb.2022.926059

Scoones I, Borras SM, Baviskar A, Edelman M, Peluso NL, Wolford W (2023) Climate change and critical agrarian studies, 1st ed. Routledge, London. https://doi.org/10.4324/9781003467960

Seneviratne SI, Nicholls N, Easterling D, Goodess CM, Kanae S, Kossin J, Luo Y, Marengo J, McInnes K, Rahimi M, Reichstein M, Sorteberg A, Vera C, Zhang X (2012) Changes in climate extremes and their impacts on the natural physical environment. In: Managing the risks of extreme events and disasters to advance climate change adaptation. A special report of working groups I and II of the intergovernmental panel on climate change (IPCC). Cambridge University Press, Cambridge, pp 109–230

Sheeja P, Singh DK, Sarangi A, Sehgal V, Iquebal M (2022) Change detection of groundwater level and quality in coastal aquifers of Malabar region in Kerala, India. Int J Environ Clim Change 19:755–768. https://doi.org/10.9734/ijecc/2022/v12i121511

Shyam SS, Geetha R (2013) Empowerment of fisherwomen in Kerala—an assessment. Indian J Fish 60(3):73–80

Shyam SS, Kripa V, Zacharia PU, Mohan A, Ambrose TV, Manjurani S (2014) Vulnerability assessment of coastal fisher households in Kerala: a climate change perspective. Indian J Fish 61(4):99–104

Singh H, Nielsen M, Greatrex H (2023) Causes, impacts, and mitigation strategies of urban pluvial floods in India: a systematic review. Int J Disast Risk Reduct 93:103751. https://doi.org/10.1016/j.ijdrr.2023.103751

Smith JB, Dickinson T, Donahue JDB, Burton I, Haites E, Klein RJT, Patwardhan A (2011) Development and climate change adaptation funding: coordination and integration. Clim Policy 11(3):987–1000. https://doi.org/10.1080/14693062.2011.582385

Society for Assistance to Fisherwomen (SAF)|Fisheries Department—Kerala. Accessed 20 March 2024. https://fisheries.kerala.gov.in/index.php/en/saf

Sreekumar S (2019) Landslide susceptibility assessment and preparedness strategies, Thiruvambaadi Grama Panchayath, Kozhikode District, Kerala. Kerala State Disaster Management Authority

Sreeraj P, Swapna P, Krishnan R, Nidheesh AG, Sandeep N (2022) Extreme sea level rise along the Indian ocean coastline: observations and 21st century projections. Environ Res Lett 17(11):114016. https://doi.org/10.1088/1748-9326/ac97f5

Surendran U, Anagha B, Gopinath G, Joseph EJ (2019) Long-term rainfall analysis towards detection of meteorological drought over Kozhikode district of Kerala. J Clim Change 5(2):23–34. https://doi.org/10.3233/JCC190010

Syamsuddin S, Marlina M, Chamzurni T, Maulidia V (2021) Indigenous rhizobacteria treatment in controlling diseases phytophthora palmivora and increasing the viability and growth of cocoa seedling. J Nat 21(2):105–113

The State of Adaptation under the United Nations Framework Convention on Climate Change (2013) Thematic report. United Nations Framework Convention on Climate Change (UNFCCC), Bonn

Topography|Kozhikode District Website|India. Accessed 5 Mar 2024. https://kozhikode.nic.in/en/about-district/topography/

United Nations Framework Convention on Climate Change. Introduction|UNFCCC. United Nations. Accessed 4 March 2024. https://unfccc.int/topics/adaptation-and-resilience/the-big-picture/introduction

United States Environment Protection Agency (2022) Temperature and precipitation. UN EPA, New York

Unnikrishnan AS, Nidheesh AG, Lengaigne M (2015) Sea-level-rise trends off the Indian coasts during the last two decades. Curr Sci 108(5):966–971

Vattarambath S (2007) Growth of tenancy movement in Malabar in the post 1921 rebellion period. Proceed Indian Hist Congr 68:654–659

Vellore RK, Deshpande N, Priya P, Singh BB, Bisht J (2020) Extreme storms. In: Assessment of climate change over the Indian region. Springer Open, Singapore, pp 155–173

Vivekanandan E (2013) Climate change: challenging the sustainability of marine fisheries and ecosystems. J Aquat Biol Fish 1(1–2):54–67

Vivekanandan E (2010) Impact of climate change in the Indian marine fisheries and the potential adaptation options. In: Meenakumari B, Boopendranath MR, Edwin L, Sankar TV, Gopal N, Ninan G (eds) Society of fisheries technologists, pp 169–185

Wabnitz CCC, Lam VWY, Reygondeau G, The LCL, Al-Abdulrazzak D, Khalfallah M, Pauly D, Deng Palomares ML, Zeller D, Cheung WWL (2018) Climate change impacts on marine biodiversity, fisheries and society in the Arabian Gulf. PLoS ONE 13(5):e0194537. https://doi.org/10.1371/journal.pone.0194537

Wolch JR, Dear M (eds) (2015) The power of geography: how territory shapes social life. In: Boston UH (ed) First issued in paperback; [Nachdr. der Ausg.]. Routledge library editions social and cultural geography, 14. Routledge, London, New York

The manufacturer's authorised representative in the EU is Springer Nature Customer Service Centre GmbH, Europaplatz 3, 69115 Heidelberg, Germany. If you have any concerns regarding our products, please contact ProductSafety@springernature.com

Printed and bound by CPI Group (UK) Ltd, Croydon, CR0 4YY
26/03/2026
02078943-0013